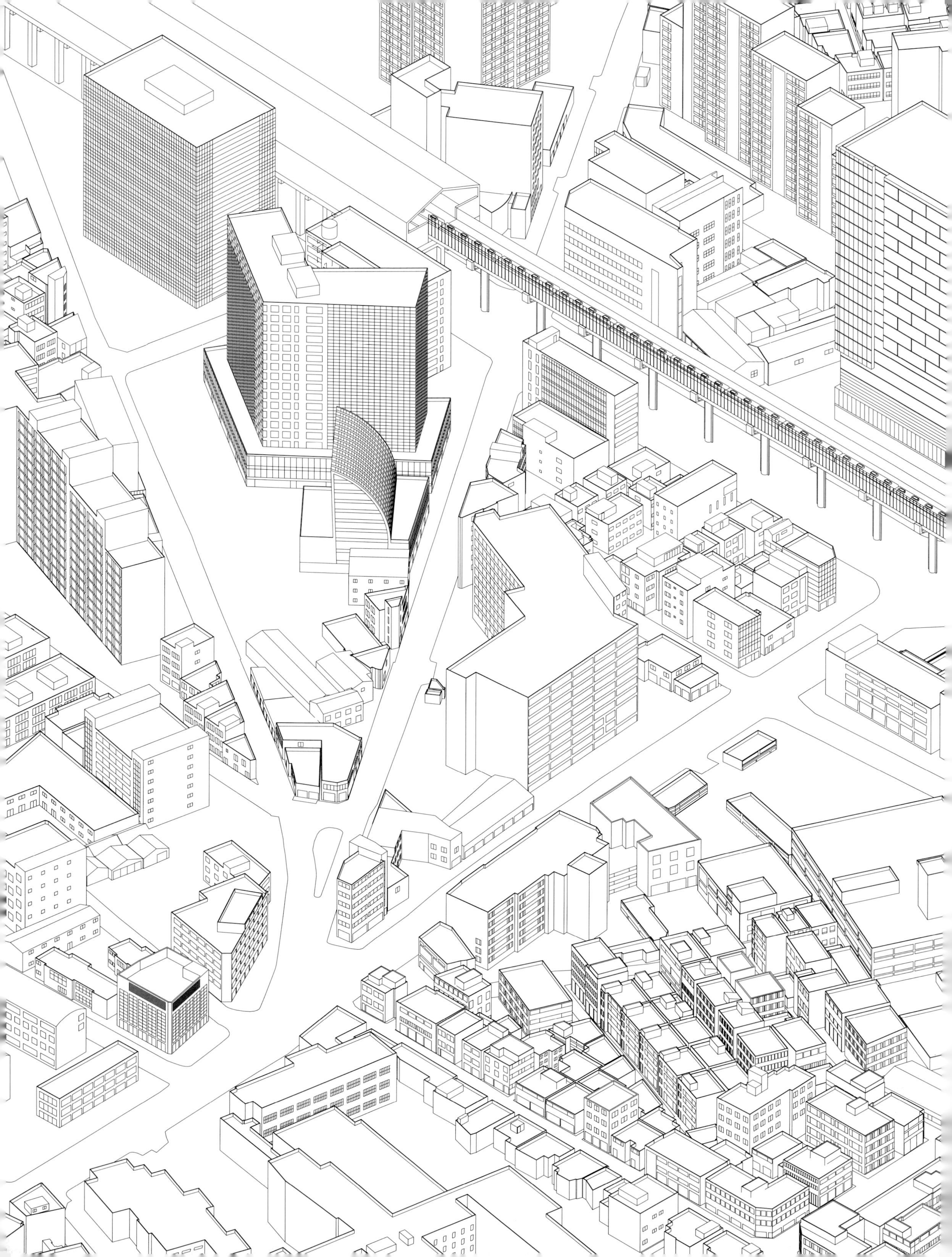

INFRASTRUCTURAL ALLIANCE
Envisioning Under-Bridge Spaces in Seongsu

Seoul National University
Department of Architecture and Architectural Engineering
John Hong Studio 2017
Sponsored by Heerim

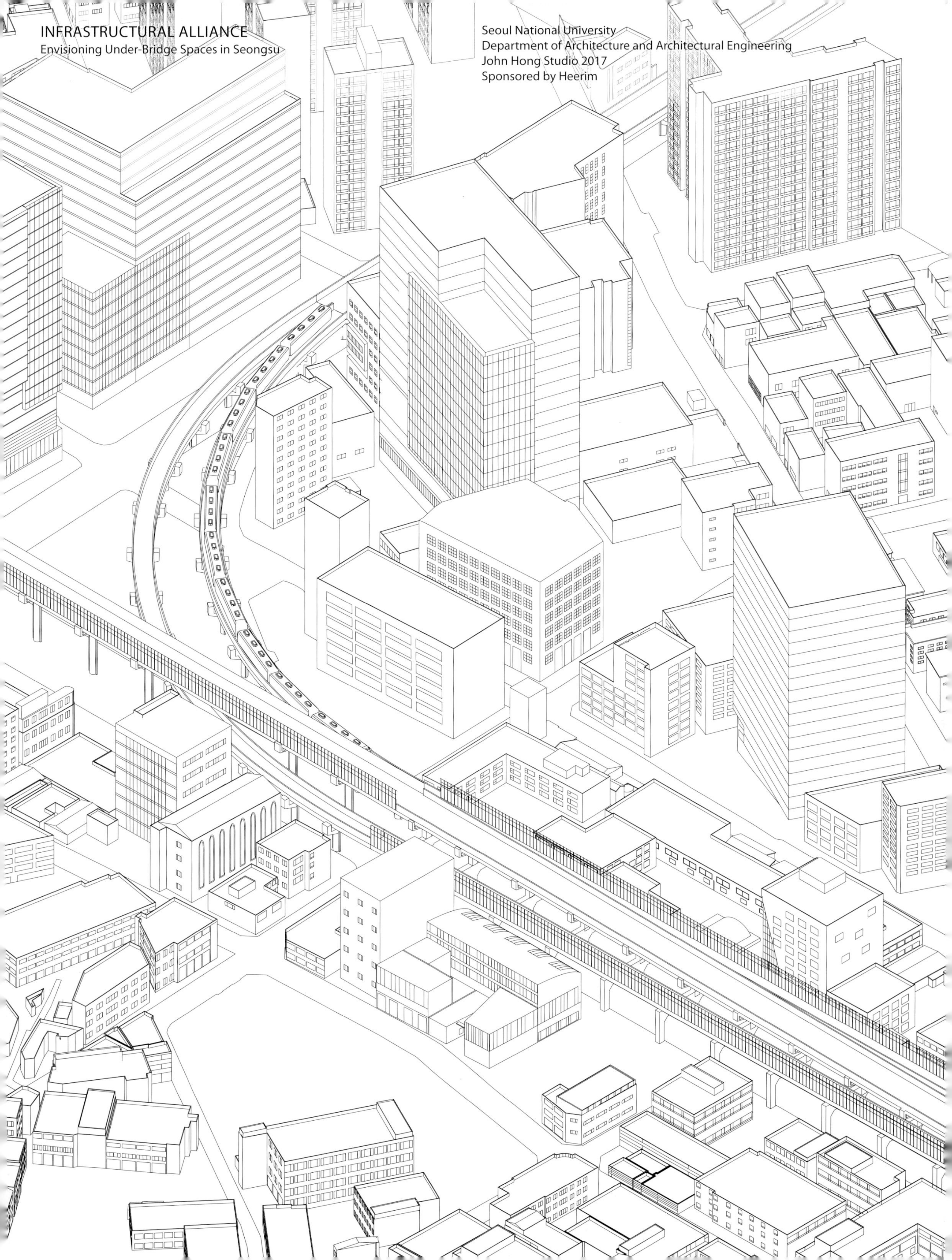

INFRASTRUCTURAL ALLIANCE

SEOUL NATIONAL UNIVERSITY
DAAE
SNU-Heerim Studio
John Hong

STUDENTS
Bonghun Jeon
Daye Kim
Ina Choi
Jinyoung Hwang
Keonhee Kim
Namhyeok Baek
Seohyun Choi
Soobin Kim
Trygve Schmidt Pedersen
Youngjoon Shim
Youngju Lee (TA)

EDITOR AND DESIGNER
Daye Kim
Ina Choi
Youngjoon Shim

This studio was made possible through the generous support of Heerim Architects & Planners.

TABLE OF CONTENTS

전방30m

Infrastructural Alliance: Envisioning Under-Bridge Spaces in Seongsu

The city of Seoul is attempting to rectify over 180 under-bridge sites. Scattered throughout the metropolitan region, these subway and traffic fly-over infrastructures cover 150 hectares of urban territories, many of which are in central areas. For a city that struggles with issues of density, these overlooked zones pose a difficult paradox: they spatially and socially divide the communities they serve.

Instead of demolishing the 'mistakes' of another era of modernity as we might have done in the past, the premise of this studio is to leverage the unique conditions and opportunities that are embedded within these forgotten spaces. The vibrant post-industrial Seongsu region is our testbed for our thesis: Through the concept Infrastructural Alliance, the goal of the studio is to link infrastructure to the emerging Seoul City policy of the '10-minute Neighborhood.' This city initiative attempts to create orient pedestrian-oriented communities around a network of social programs custom-tailored to the neighborhoods they serve. Even as these under-bridge spaces have been neglected, the very condition that they remain unbuilt and adjacent to the city's transport systems, imbues them with the potential to become central community nodes.

Although the '10-minute Neighborhood' is a powerful metaphor, it remains undefined, especially in the context of Seoul. In such a large metropolitan region, community scale aggregations are necessary but are challenged by infrastructural disconnections, piecemeal urban renewal, and complex but rich historical layers. To structure our interpretation, we create our own theses through a reading of two radically different versions of the neighborhood represented in Josep Lluís Sert's Neighborhood Unit Plan and Michel De Certeau's Walking in the City. The former text brings to light a convincing rationality that some critics have nonetheless argued requires aggressive inner-city urban renewal to make the idealizations work. The latter text argues that the city, in its inherent density, resists a totalized understanding and dissolves into hallucinogenic layers of memory and subjectivity understood in fragments. If for a moment we reduce these logics to 'top-down' and 'bottom-up,' a solution for Seongsu's infrastructure lies somewhere between the poles of comprehensive planning and open-ended systems of citizen engagement.

서울시는 총 180개소가 넘는 고가하부 영역에 대한 재생을 시도하고 있다. 대도시권 전역에 분산되어있는 지상철 및 고가도로는 150헥타르에 달하는 공간을 차지하고 있으며, 이 중 대부분이 도심부에 존재한다. 간과해버리기 쉬운 이와 같은 공간은 역설적 상황을 만들어내고 있다: 고밀화 문제에 직면하고 있는 도시에서 서비스를 제공해야 할 커뮤니티를 되려 공간적, 사회적으로 분리하는 것이다.

소위 모더니즘 시대에 행해진 '실수' 들을 부수어 없애는 대신 이러한 잊혀진 공간들이 가진 고유한 상황과 가능성을 함양하는 것이 스튜디오의 전제이다. 그리고 산업 부흥기 이후로 활발히 변화하고 있는 성수동 일대가 그러한 생각을 펼칠 실험대상으로 상정되었다. 기반 연합이라는 기조하에, 스튜디오는 서울시가 근래 제시한 '10분 동네' 정책과 성수의 기반 시설을 융합하는 것을 목표로 한다. 현재 서울시는 개별 동네가 가진 고유한 상황들 속에서 사회적 관계망을 도출하여 이를 중심으로 보행 친화적인 커뮤니티 형성을 꾀하고 있다. 고가 하부는 오랫동안 경시되었지만, 도시의 교통 체계와 인접한 동시에 공지로 남아있다는 점에서 커뮤니티의 중심지로서의 가능성을 품고 있다.

'10분 동네'라는 개념이 강력한 은유임에도, 서울이라는 특수한 상황 속에서는 그 의미가 모호해질 수밖에 없다. 서울과 같은 규모의 거대한 대도시에서는 커뮤니티 단위의 사회 활동이 필수적이다. 하지만 기반 시설이 일으키는 분절, 단편적인 도심 재개발과 복잡하면서도 풍부한 역사적 층위/맥락 등이 도시 내 커뮤니티 활동의 존속을 위협하고 있다. 이에 대하여 호세 루이스 서트의 Neighborhood Unit Plan과 미셸 드 세르토의 Walking in the City라는 근린주구에 대한 상반적 견해를 밝히고 있는 두 저서를 통해 스튜디오만의 논지를 구축하였다. 전자는 몇몇 비평가들에게 논쟁의 대상이었던 설득력 있는 합리성을 강조하는데, 그와 동시에 이상적인 작동 구조를 만들어내기 위해서 적극적인 도심부 내 재개발을 긍정하고 있다. 후자는 도시란 그 고유의 밀도와 특성 때문에 전체주의적 이해가 불가능하며, 그 대신 기억과 주관이 녹아든 환영과 같은 레이어들에 대한 부분적 이해만이 가능하다고 주장한다. 이 두 주장을 단순화하여 각각 하향식 계획과 상향식 계획으로 이해한다면, 성수의 기반 시설에 대한 해법은 총괄적인 계획을 통한 해결과 시민 참여를 수반하는 열린 결말식 해결이라는 양극단 사이 어딘가에 자리할 것이다.

LEATHER ZONE
바우레더
461-5613
동광피혁
T.465-5506
SHINE SHOE
T.497-8590
성수
463-1888
성수
부동산
레지던스

정
밀
469-0808
성수이로7길 46→1
Seongsui-ro 7-gil
현대건설
성수역 테라스타워
1599-8864
상가분양
(커피숍,편의점,부동산
분식점,음식점)
디자인
견인지역
일방통행

P15
P16
P17
79구 4668

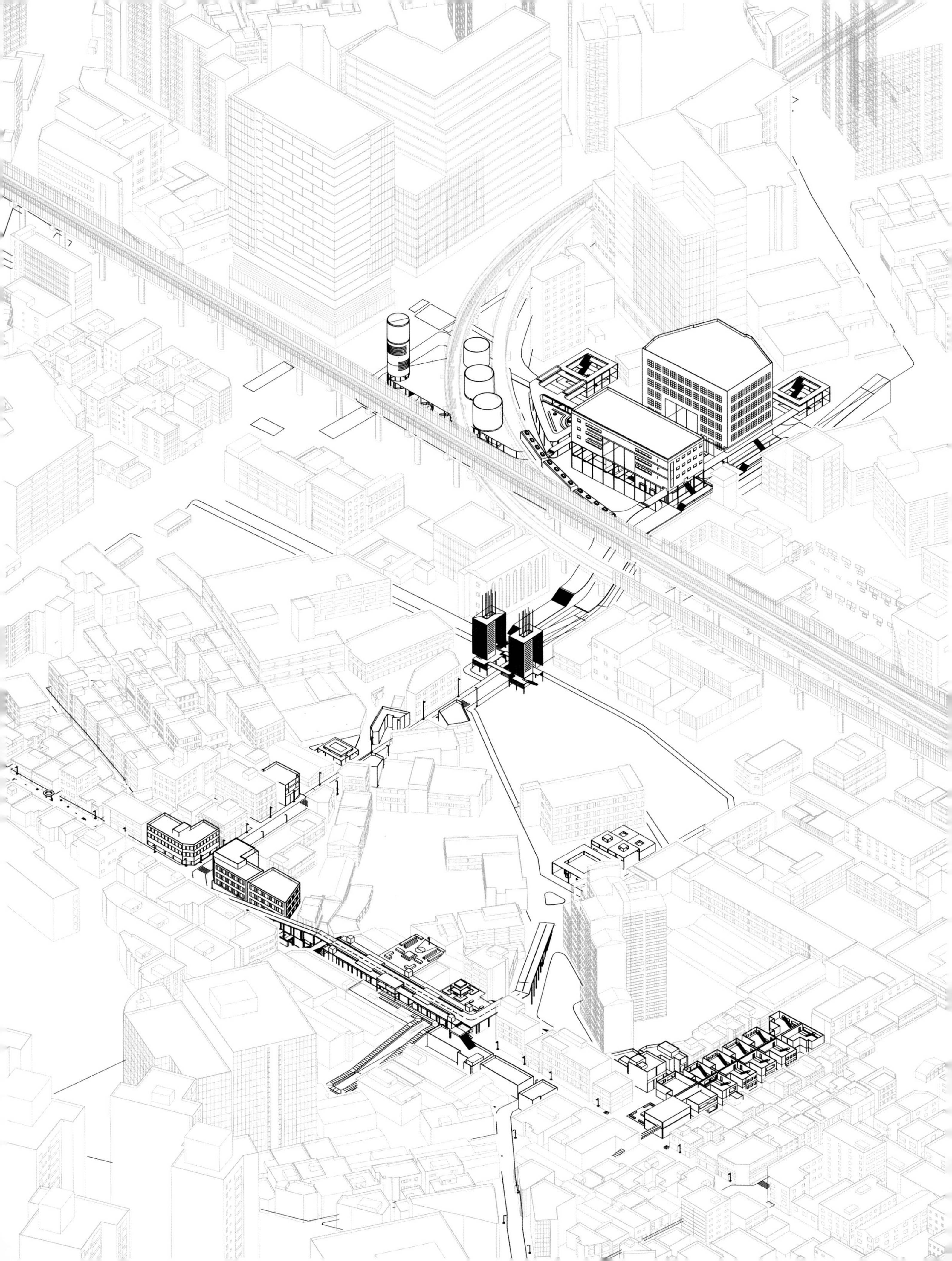

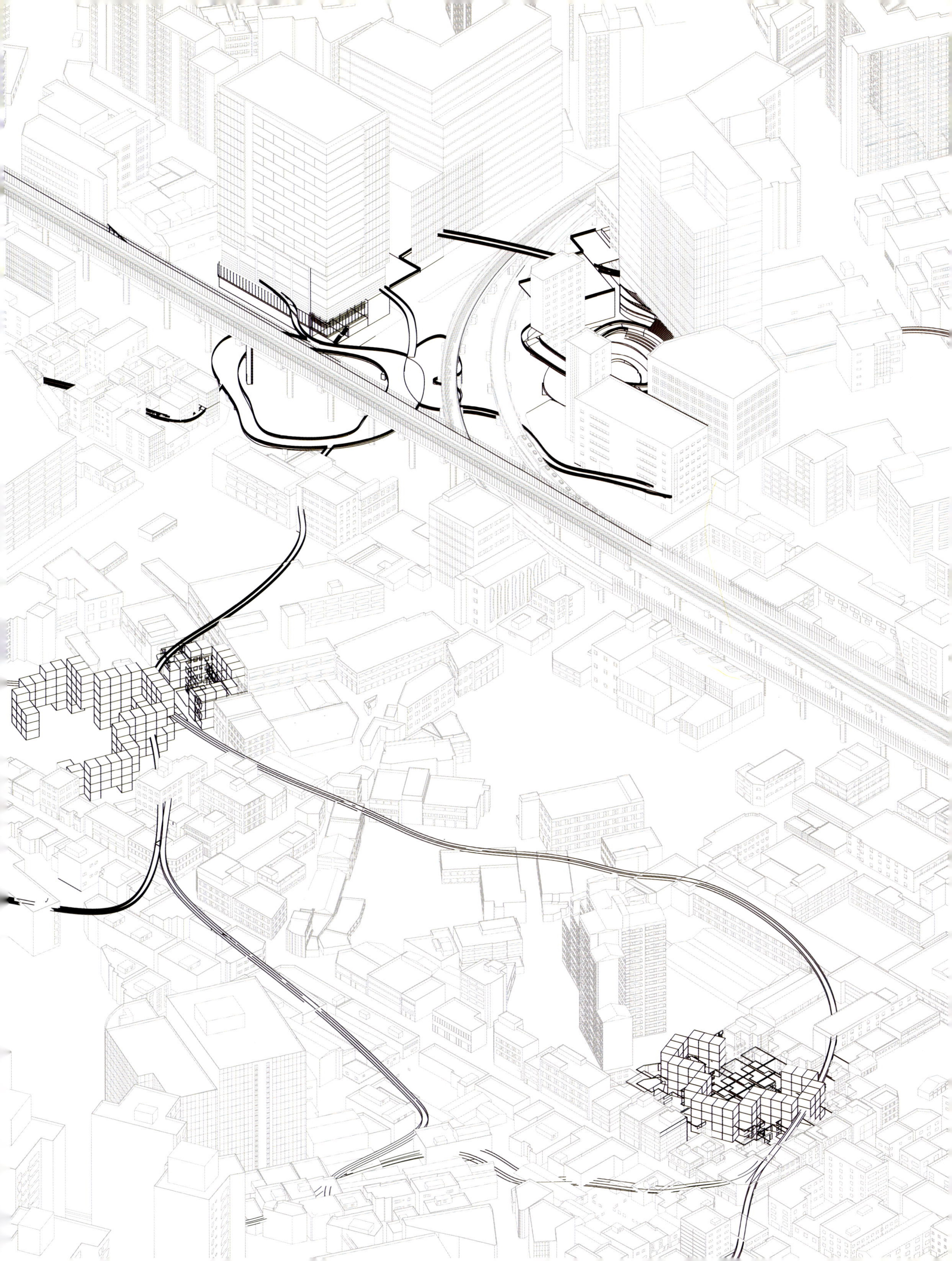

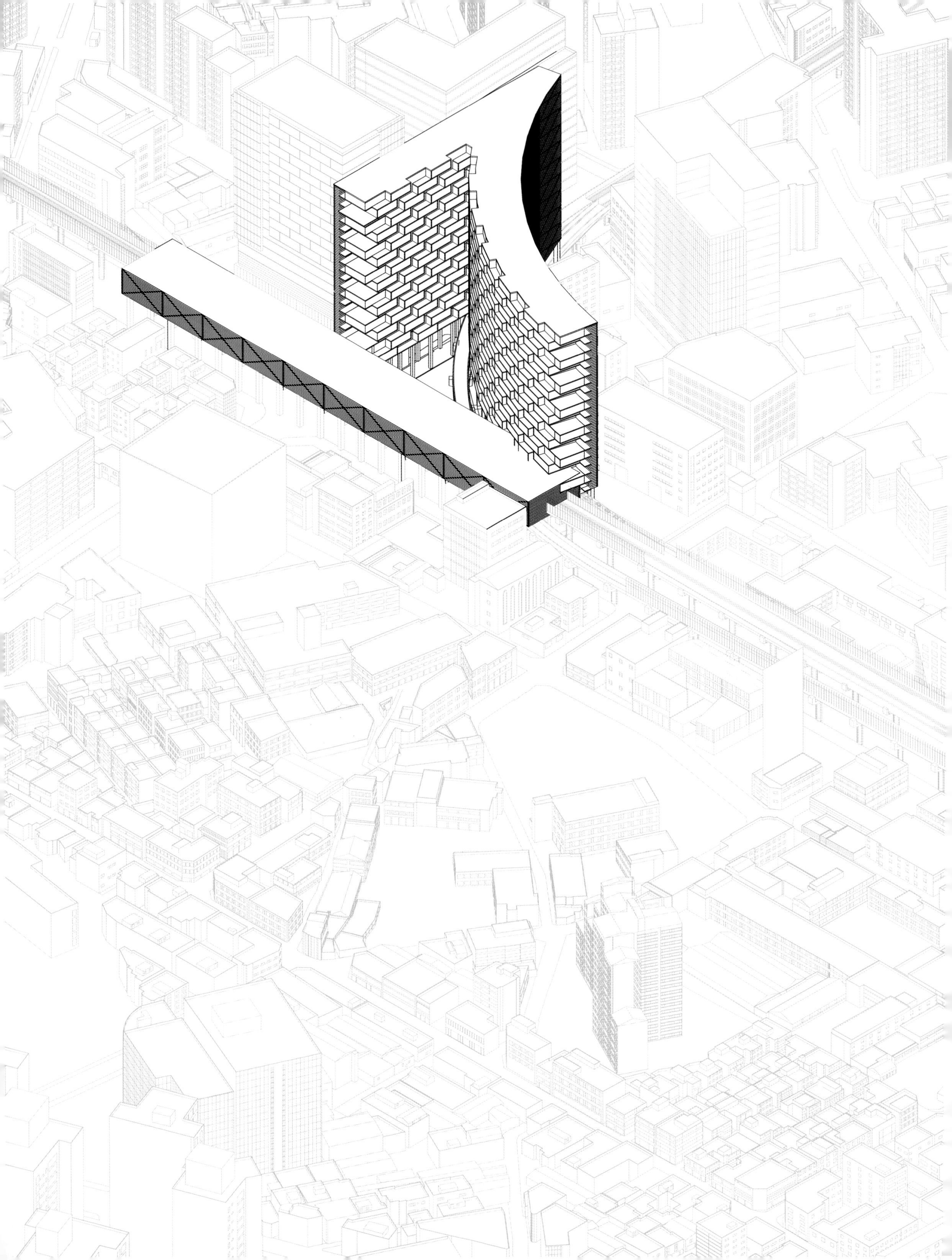

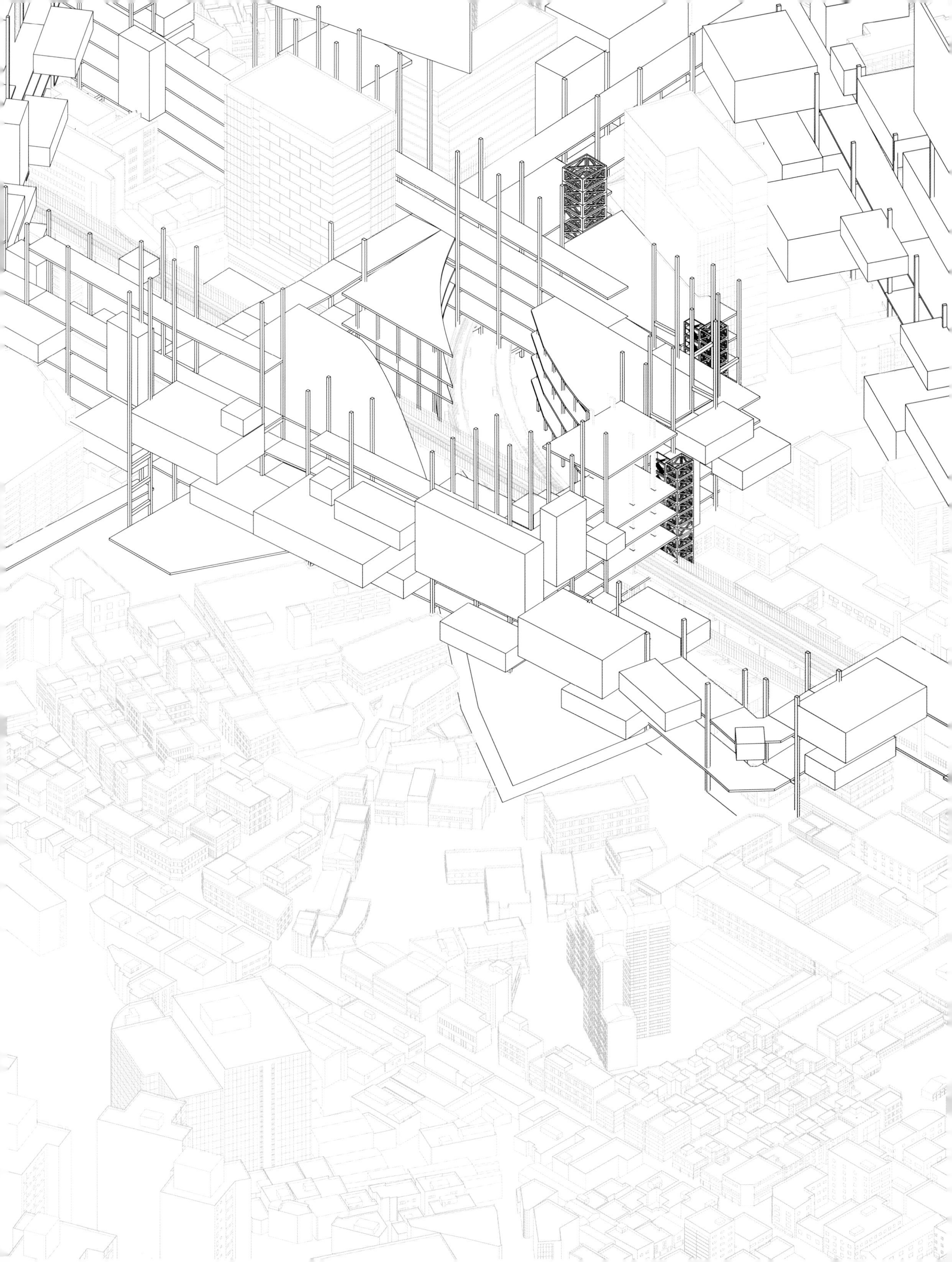

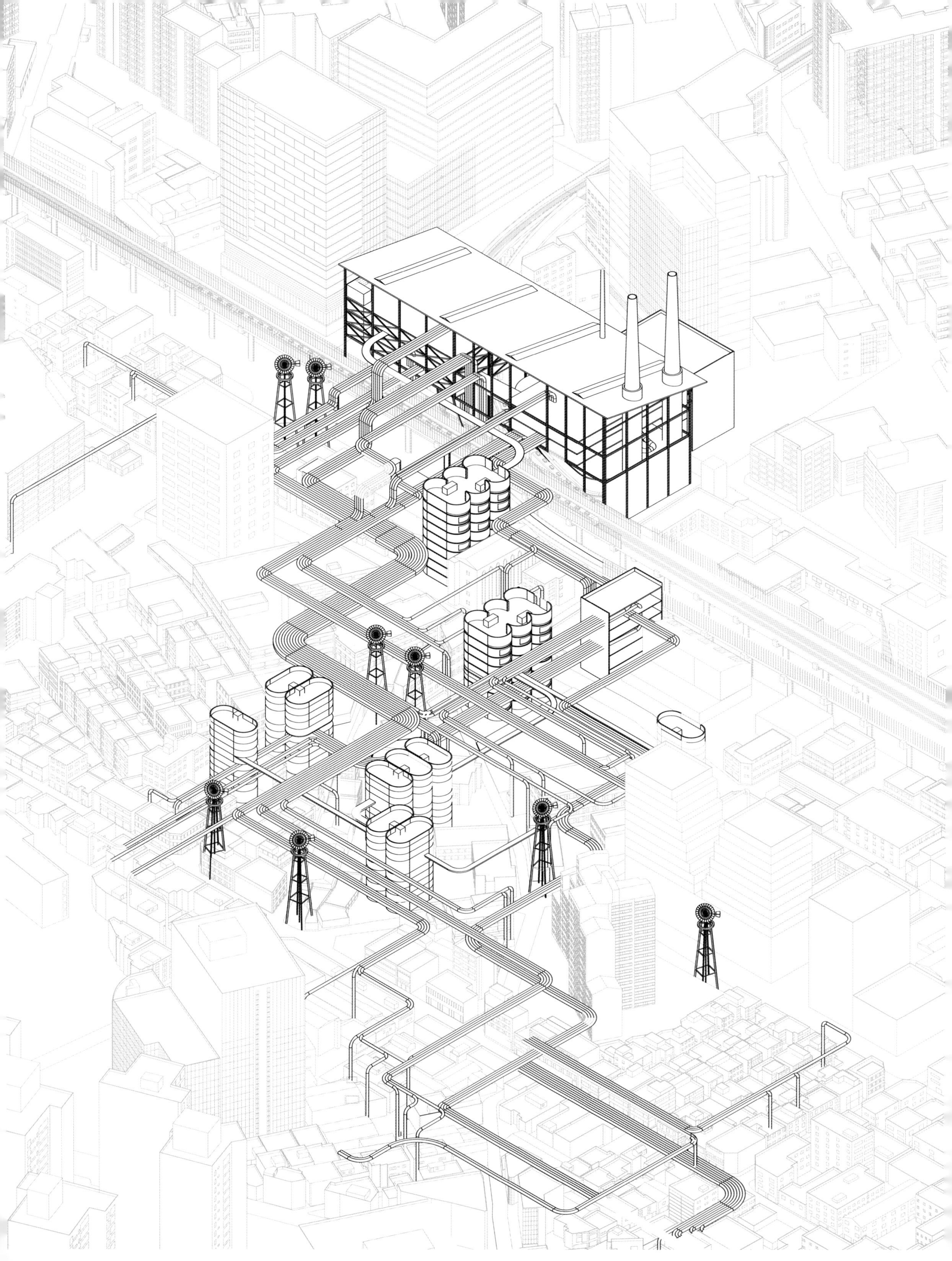

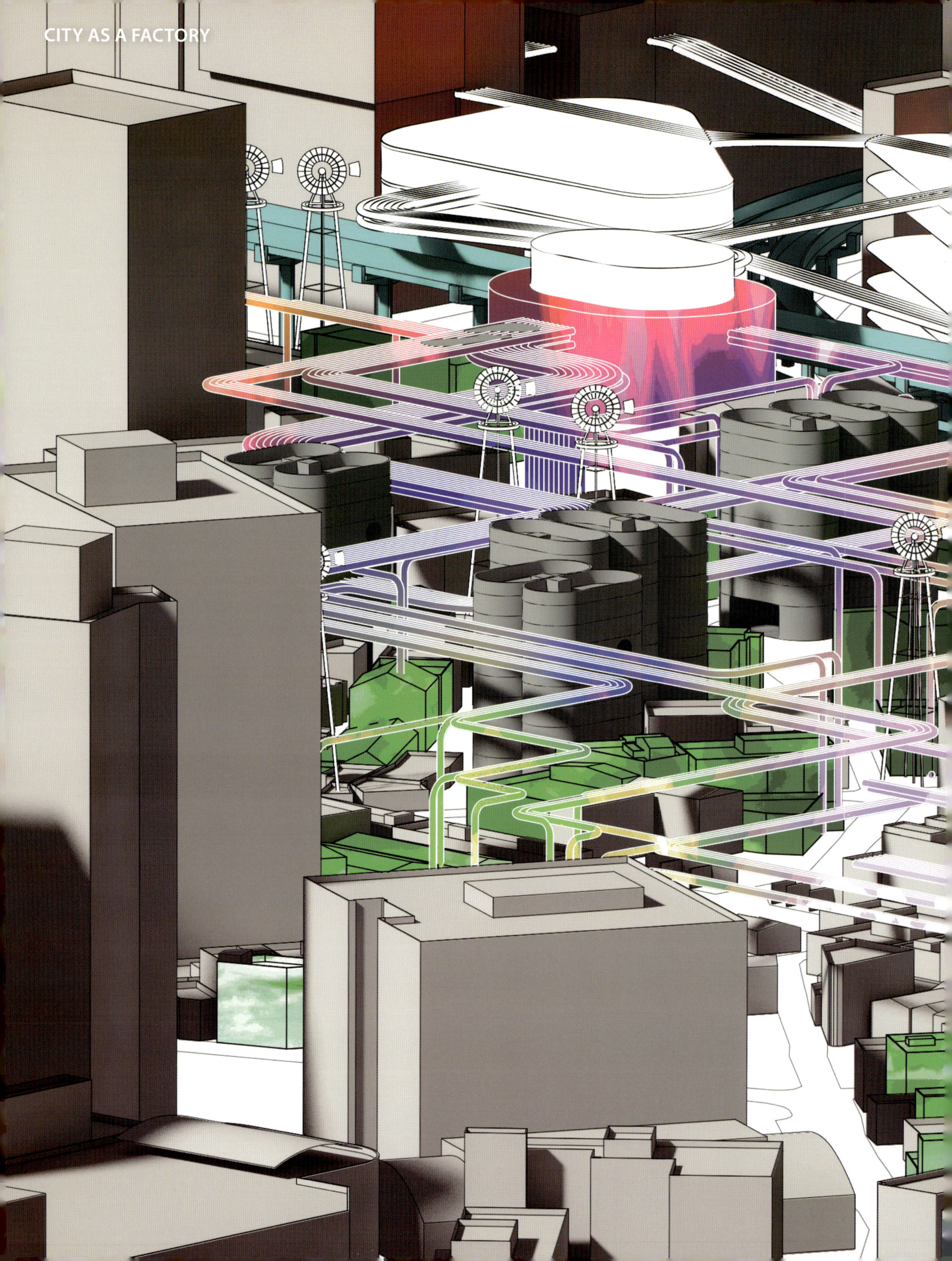

CITY AS A FACTORY

Seohyun Choi

This project creates an infrastructure through dealing with the resulting materials of Seongsu's manufacturing industry. The pipes collect and redistribute every material or byproduct according to their own distinctive character. It also distributes renewable energy to the businesses forming a new vital infrastructure for the future industry of Seongsu.

1

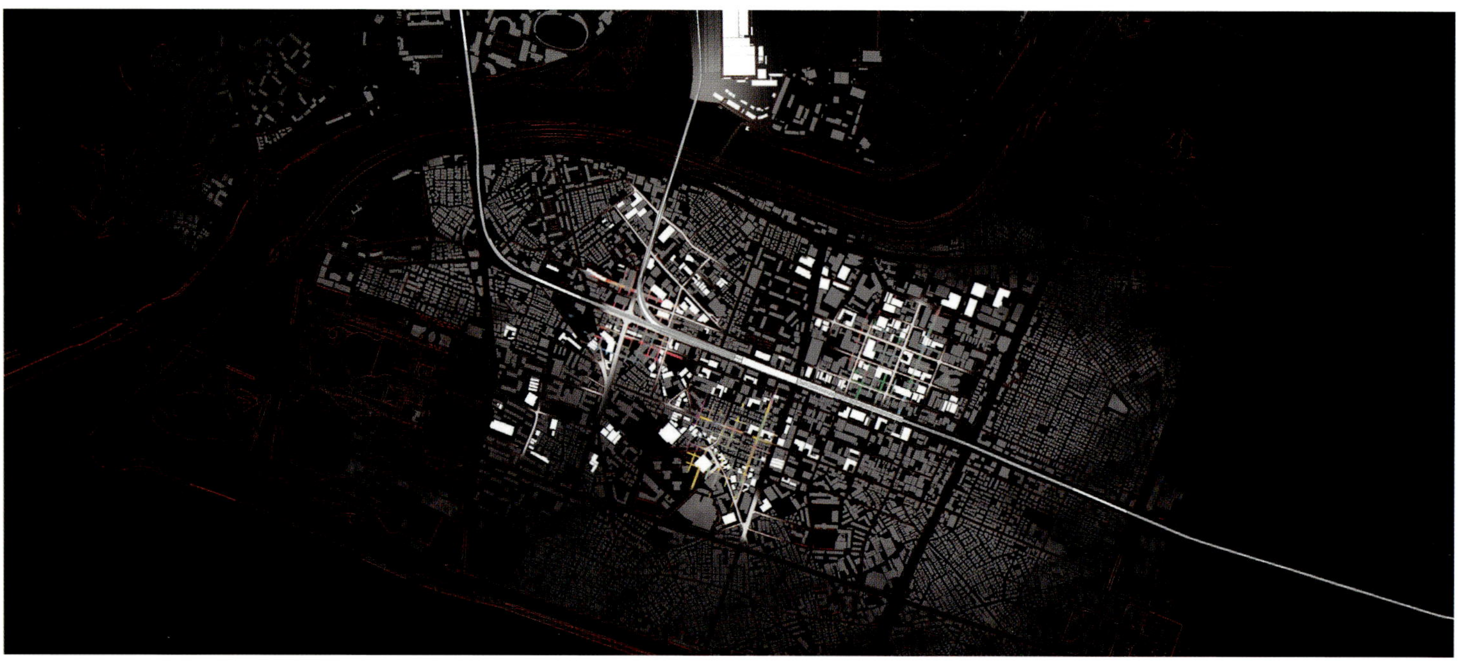

2

1. Section Perspective - Main Factory

2. Masterplan Connection
The pipes support not only small-scale manufacturing plants but also the start-up companies who can make the most out of the recyclable materials. A better working environment and pedestrian-oriented urban condition are expected accordingly.

3. Unit Section

4. Logistics Diagram

5. Program and Circulation Diagram

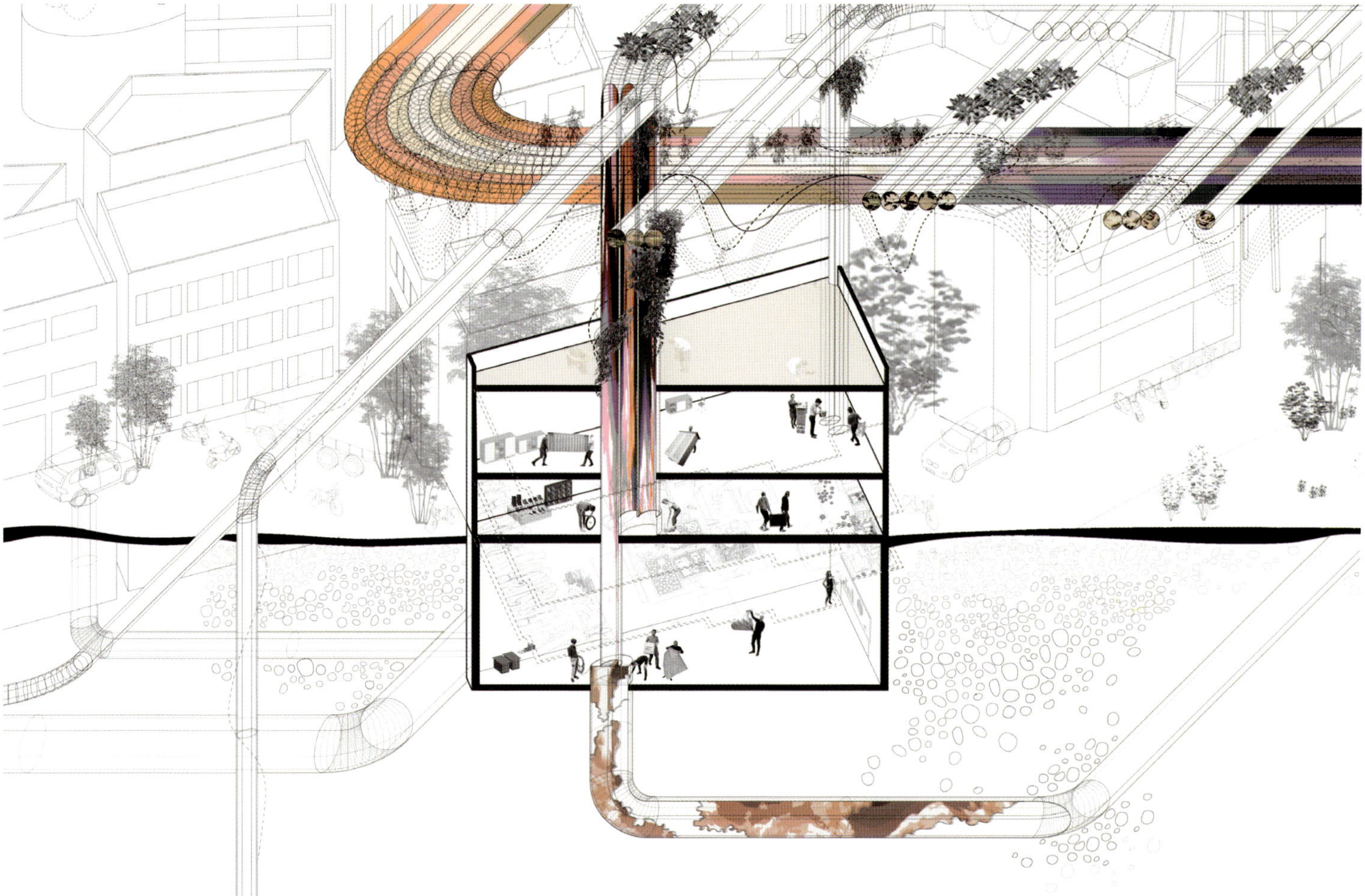

3

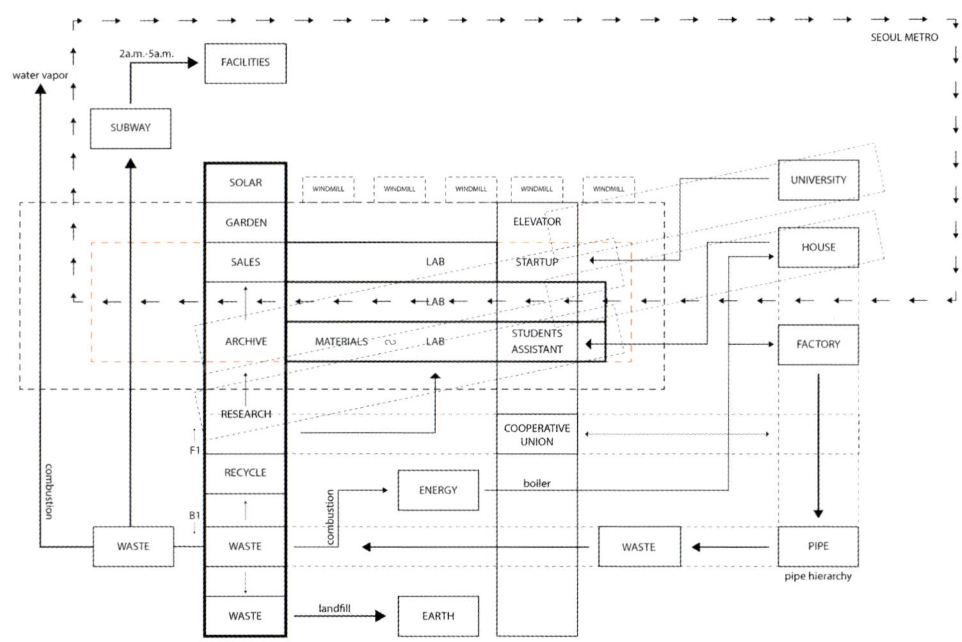

4

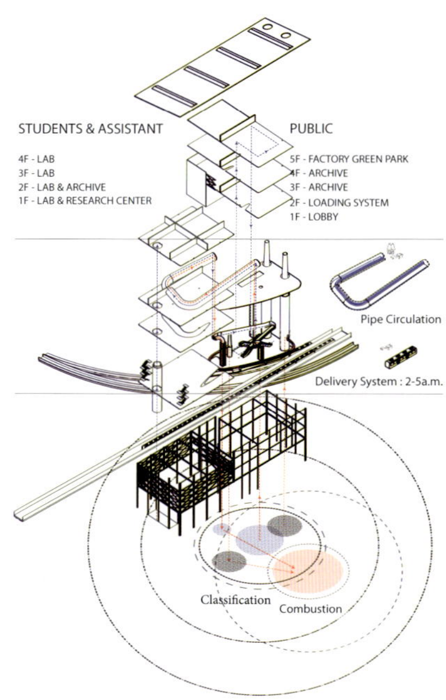

5

6. View from Yeonmujang-gil

7-9. Plans
The pipes converge at the Main Factory, located at the junction of the 2 overpasses. Pollutants from the regional factories, previously unchecked, are recollected and purified underground, while the remaining materials are once again classified.

6

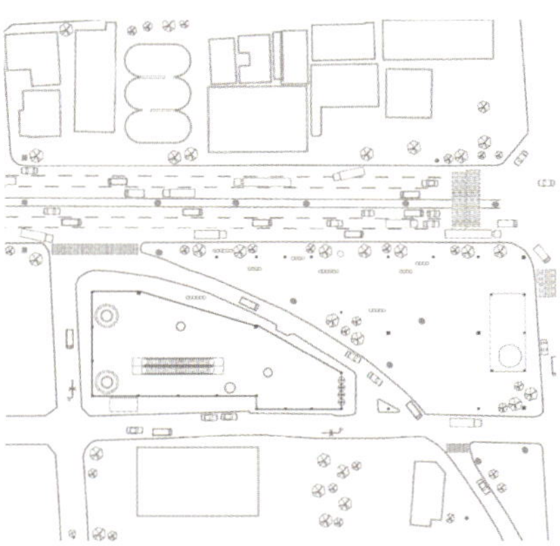

7. 1st Floor Plan

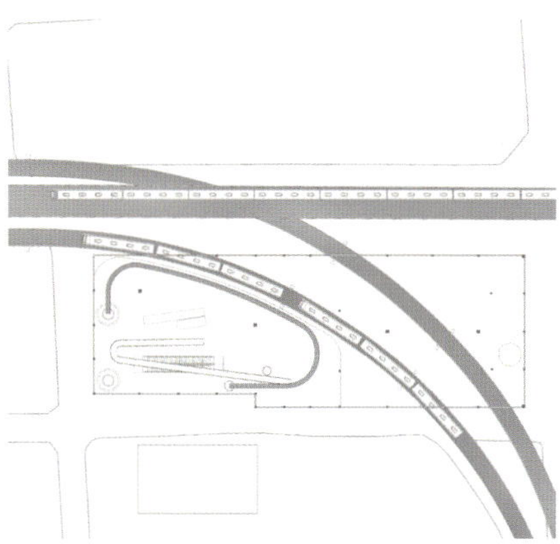

8. 2nd Floor Plan

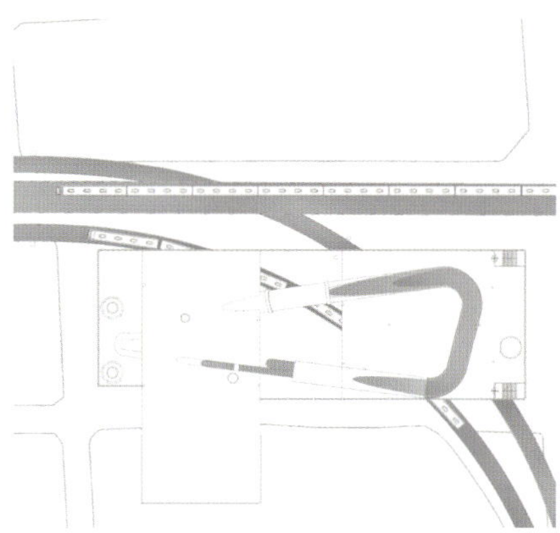

9. 3rd Floor Plan

10. View inside Main Factory
The remaining reusables are sent to the start-up's laboratory on the ground floor, while the non-recyclables are moved up to the upper subway deck. Making use of subway transportation's regular closure from 2am to 5am, these materials are sent to the new disposal plant just north of Seongsu, or moved down to the basement to be incinerated, supporting the nearby area in the form of thermal energy. This makes the Main Factory a purgatory space for industrial materials, a bridge transpiercing the industrial history of Seongsu.

11-12. Model View

10

11

12

INDUSTRIAL PLAYGROUND FOR NEW GENERATION

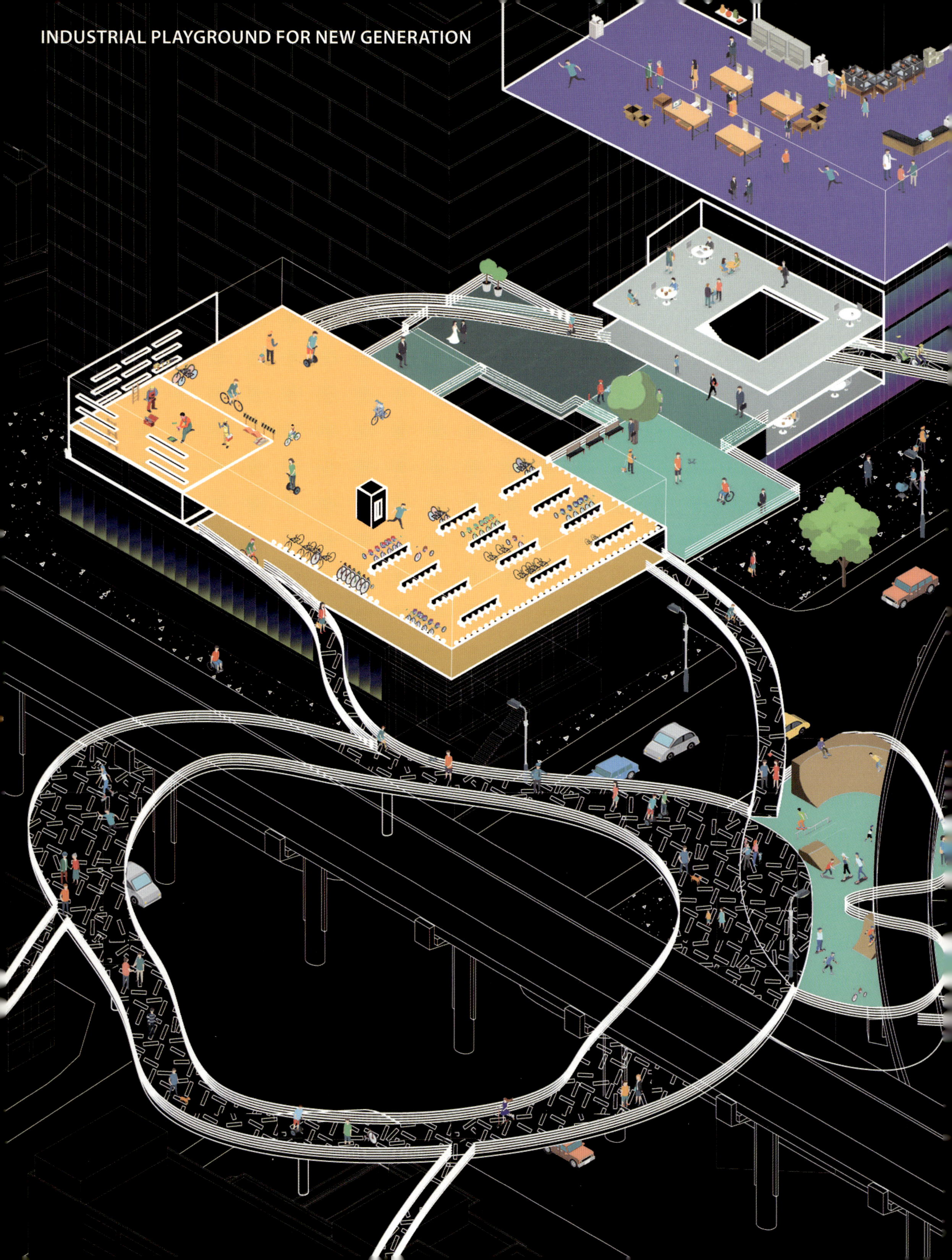

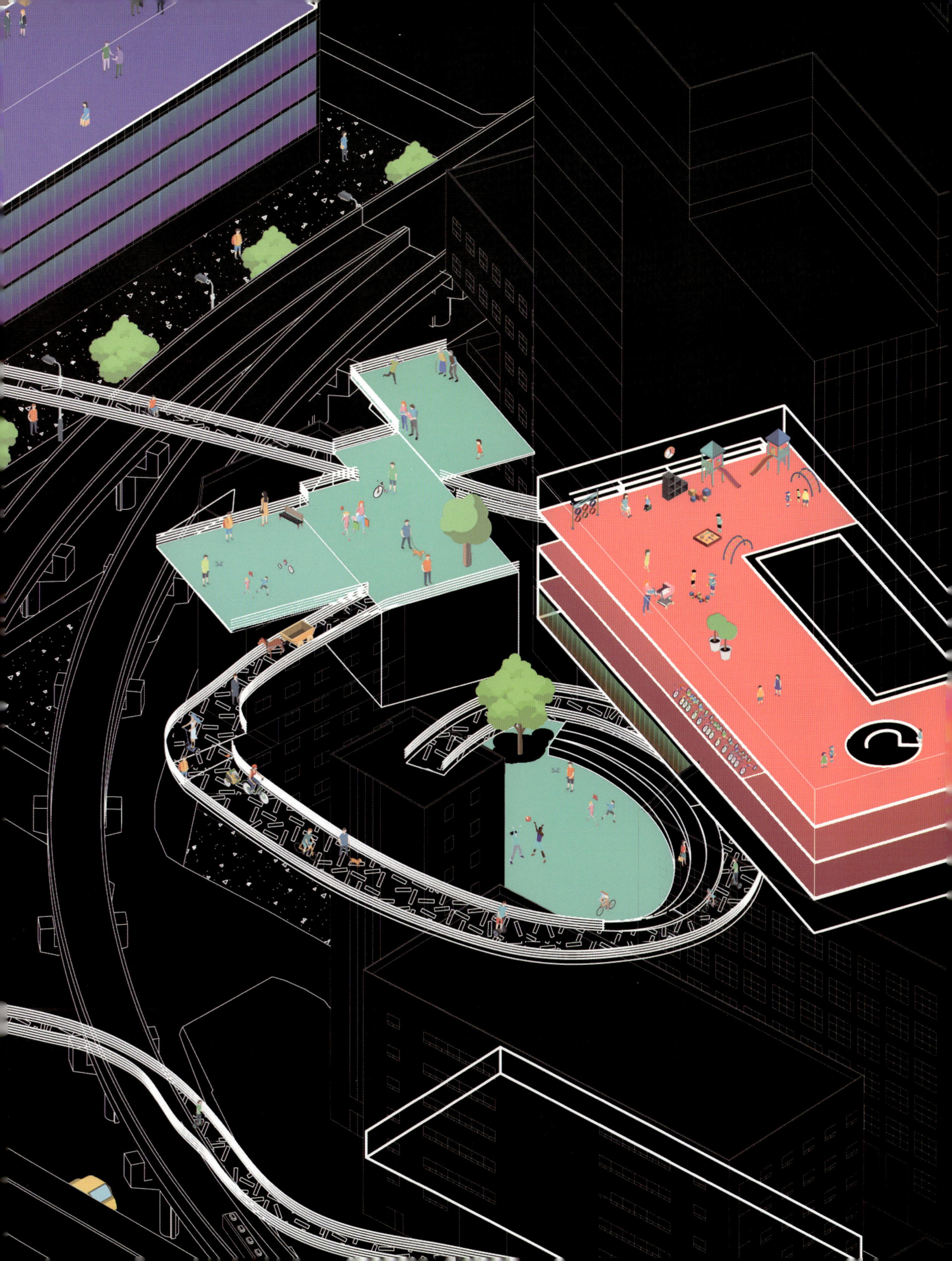

INDUSTRIAL PLAYGROUND FOR NEW GENERATION

Jinyoung Hwang

This project discusses social segregation happening in many layers, in industry, between generation and within every day urban life. The project opposes the traditional distinction between work and life, and by pointing out the unseen loss of vitality that results from the gap, emphasizes the need of cooperative space.

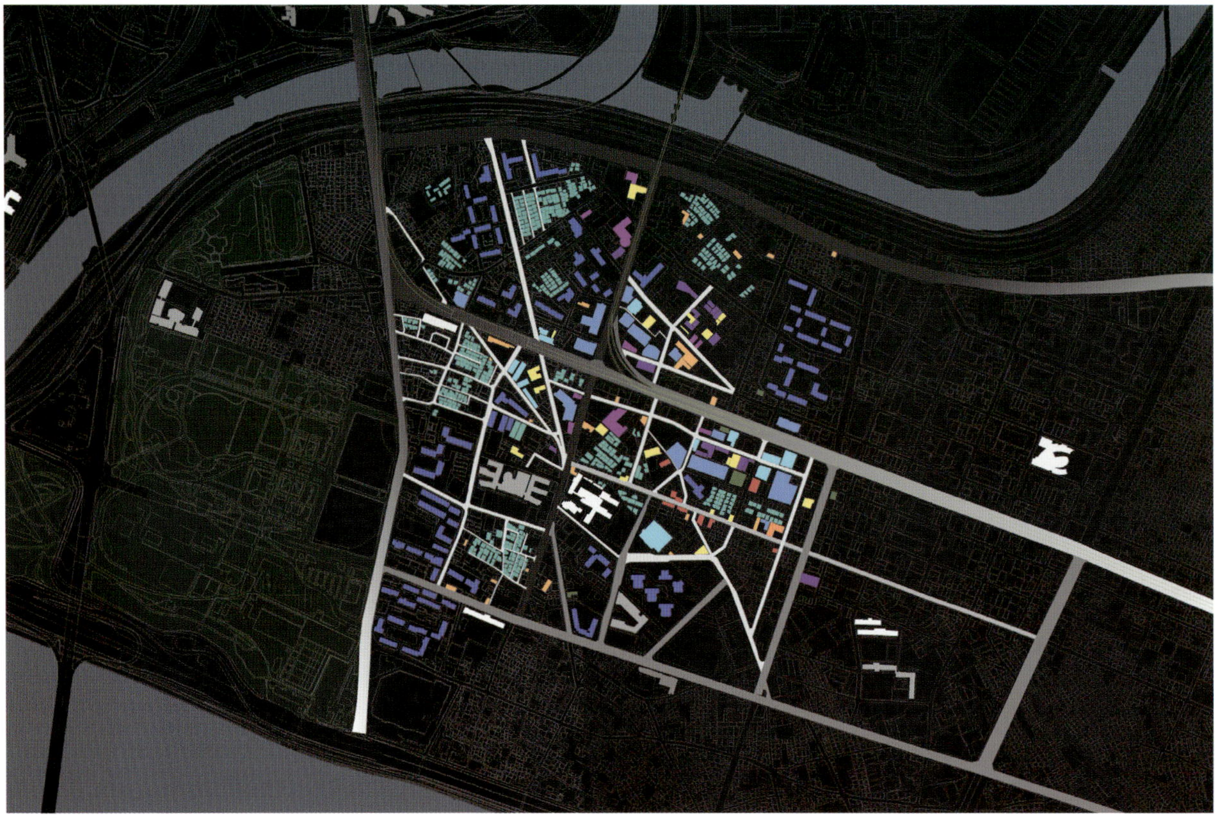

1

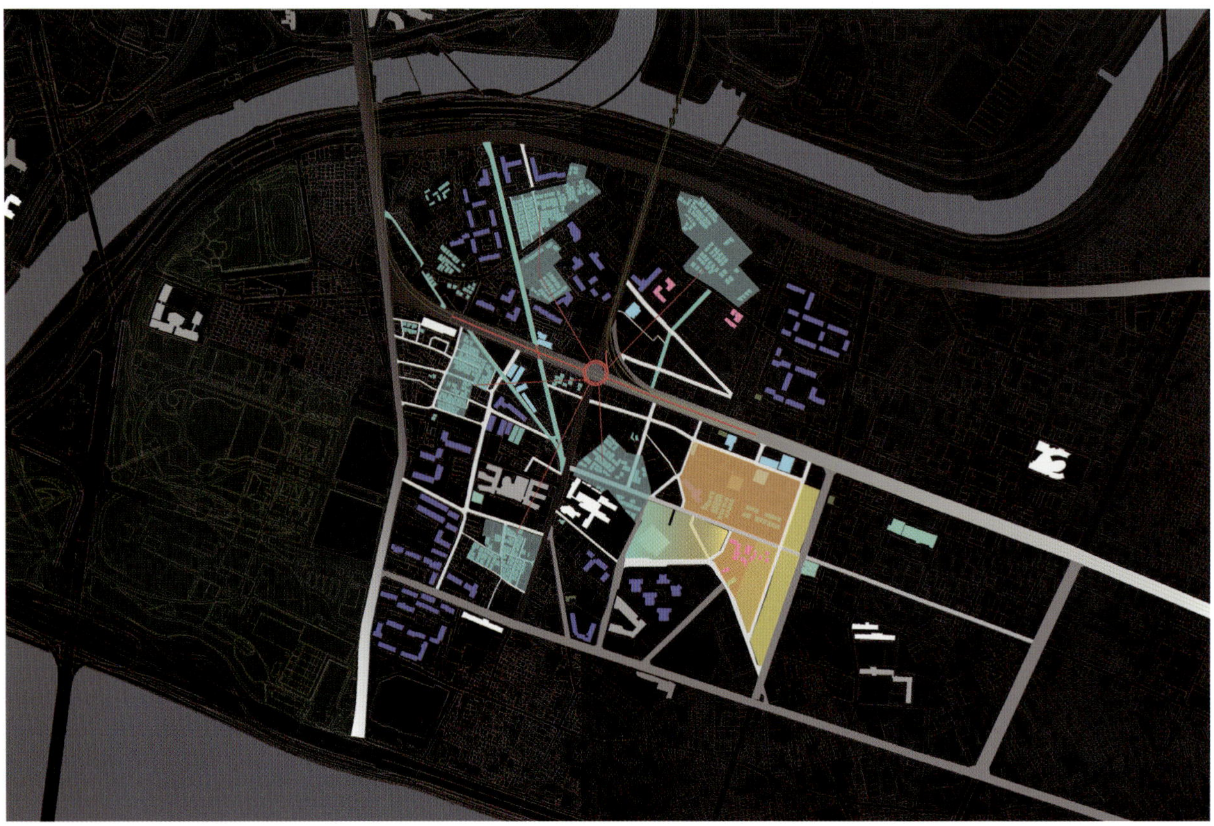

2

1-2. Masterplan Diagram Before/After

3. Circulation for Each Floor

4. Main Atrium as One of the Cooperative Spaces

Furthermore, lifelong education is introduced to reduce the disparity among generations, in hope for a more cohesive neighborhood. These approaches will not only hold generations altogether, but also afford Seongsu a new urban existence, where life itself becomes a work and work becomes a life.

5. East-West Section Perspective

6. North-South Section Perspective

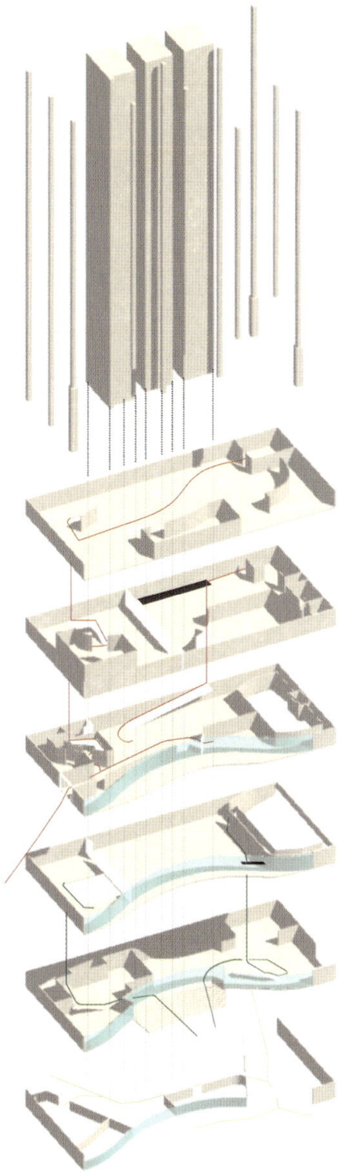

3

4

5

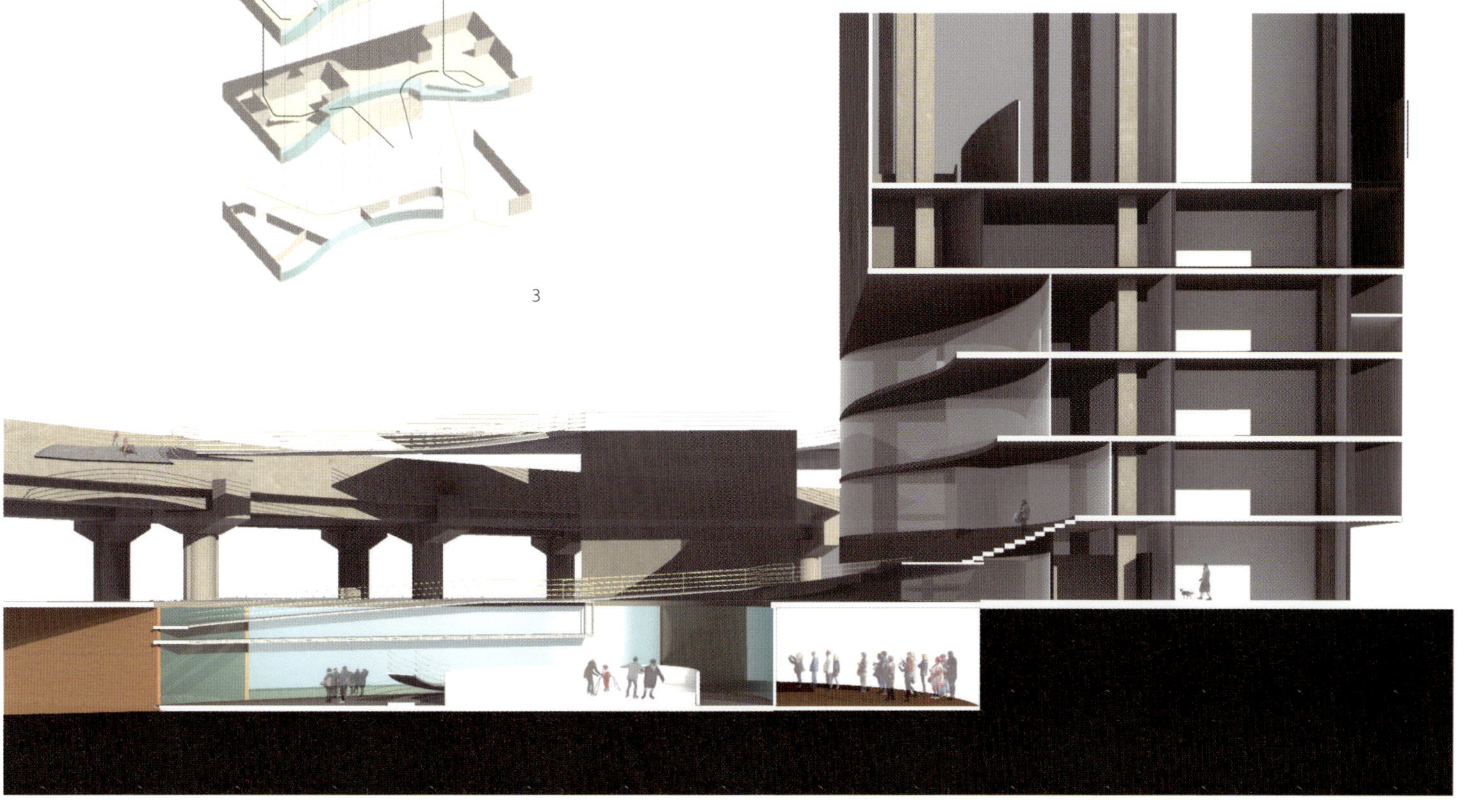

6

7-8. Programs
Underused factories in Seongsu are assigned new programs and then spatially connected to one another, either by penetrating overpasses or green space in between. The programs are provided through the new daily existence and include facilities for bike commuting, daycare, learning and resting.

9-12. Plans
Resulting activity and movement will revitalize Seongsu by creating a new identity, and nearby attractions will also be provided convenient accessibility. But these are not so much the final aim as a result : the former pedestrian-oriented connections are more than the sum of the coexistence of industry and dwelling.

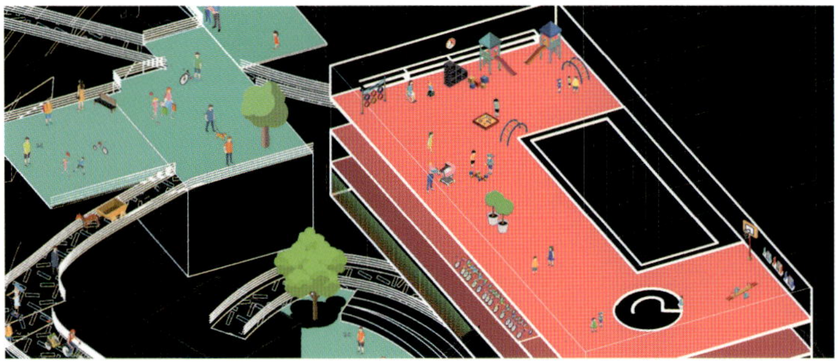

7

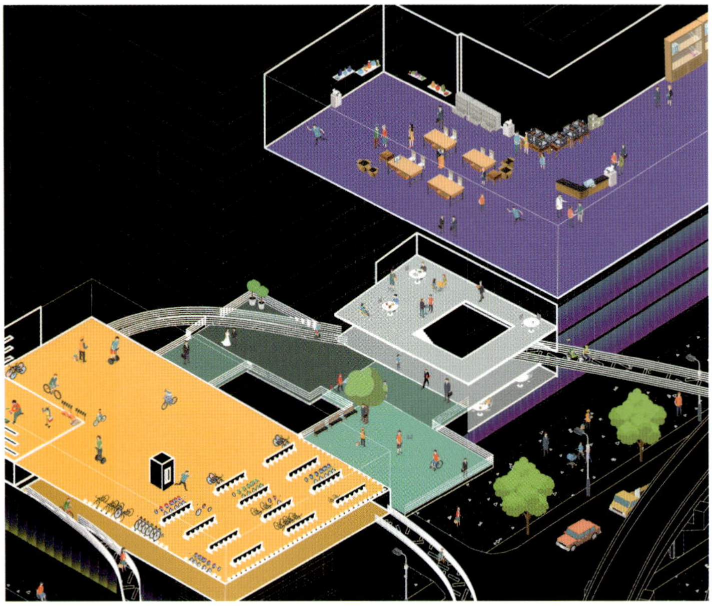

8

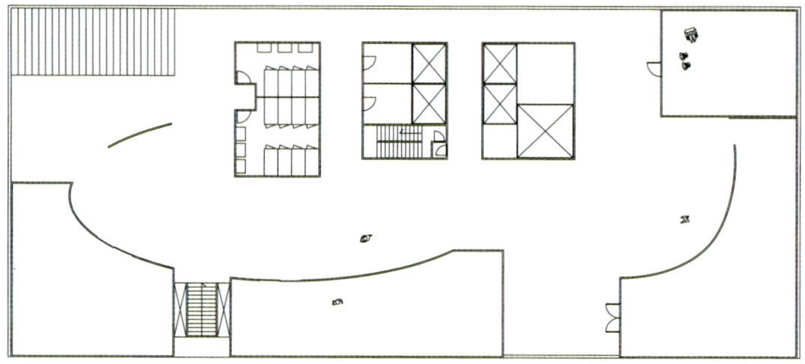

9. 6th floor

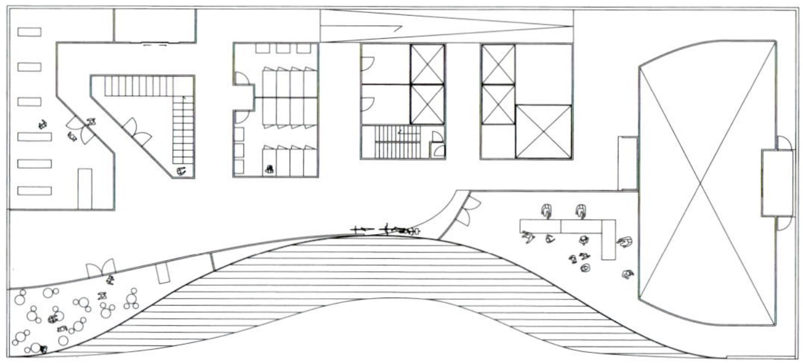

10. 4th floor

11. 2nd floor

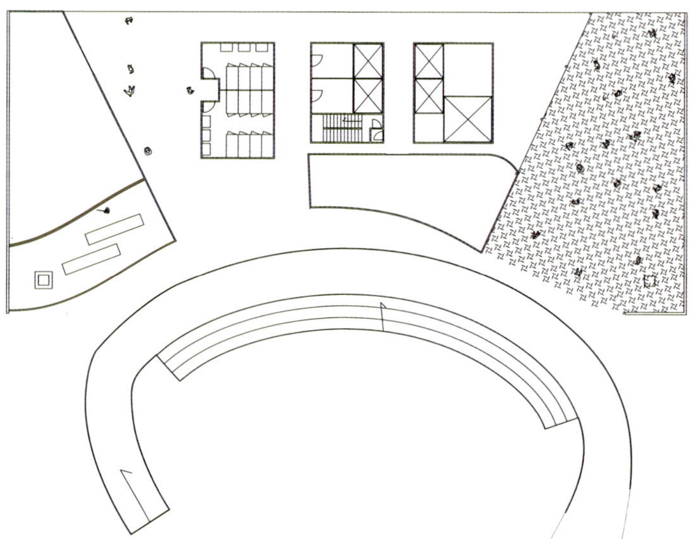

12. Ground floor

13-14. Model shot

13

14

SUBWAY DISTRIBUTION HUB IN SEONGSU

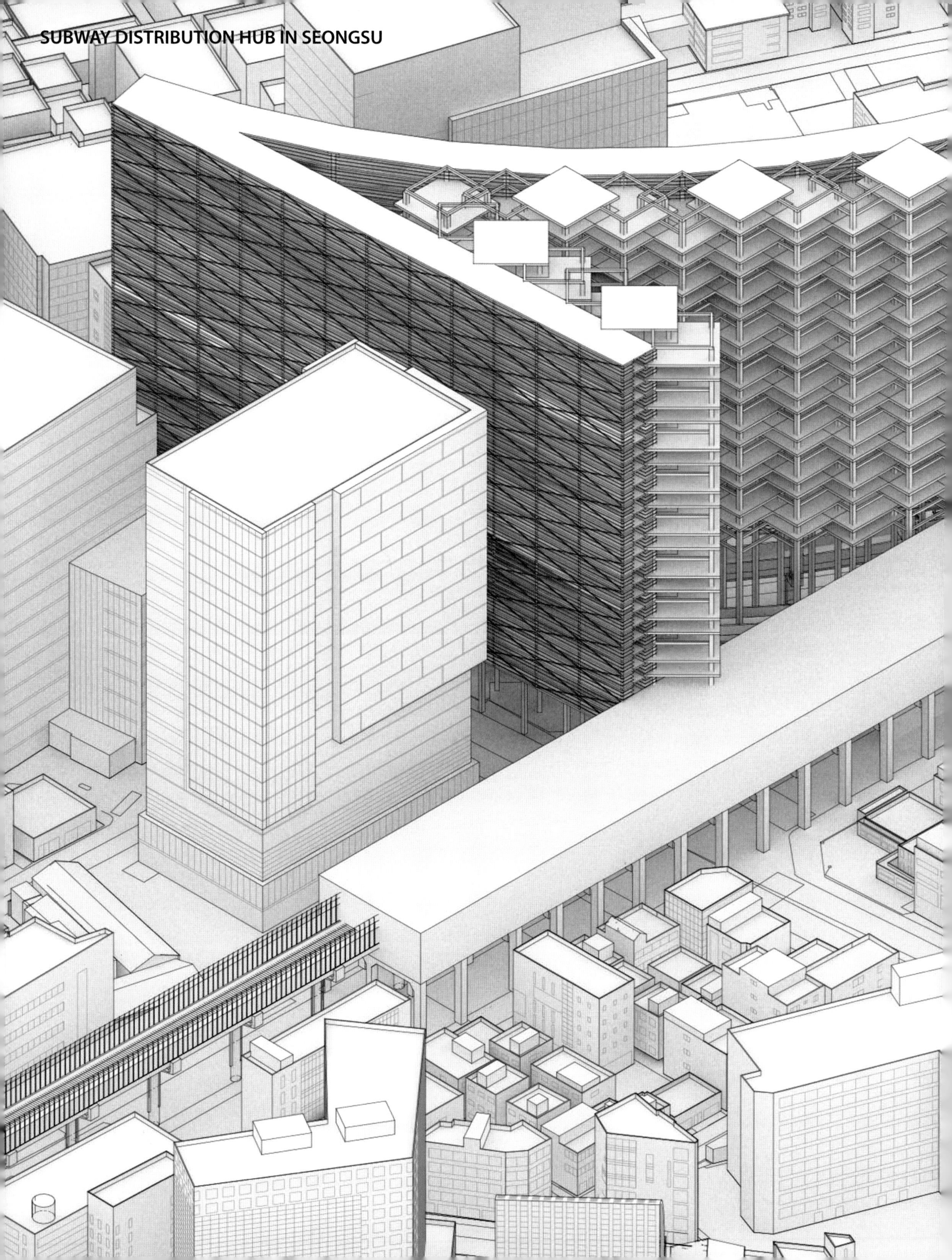

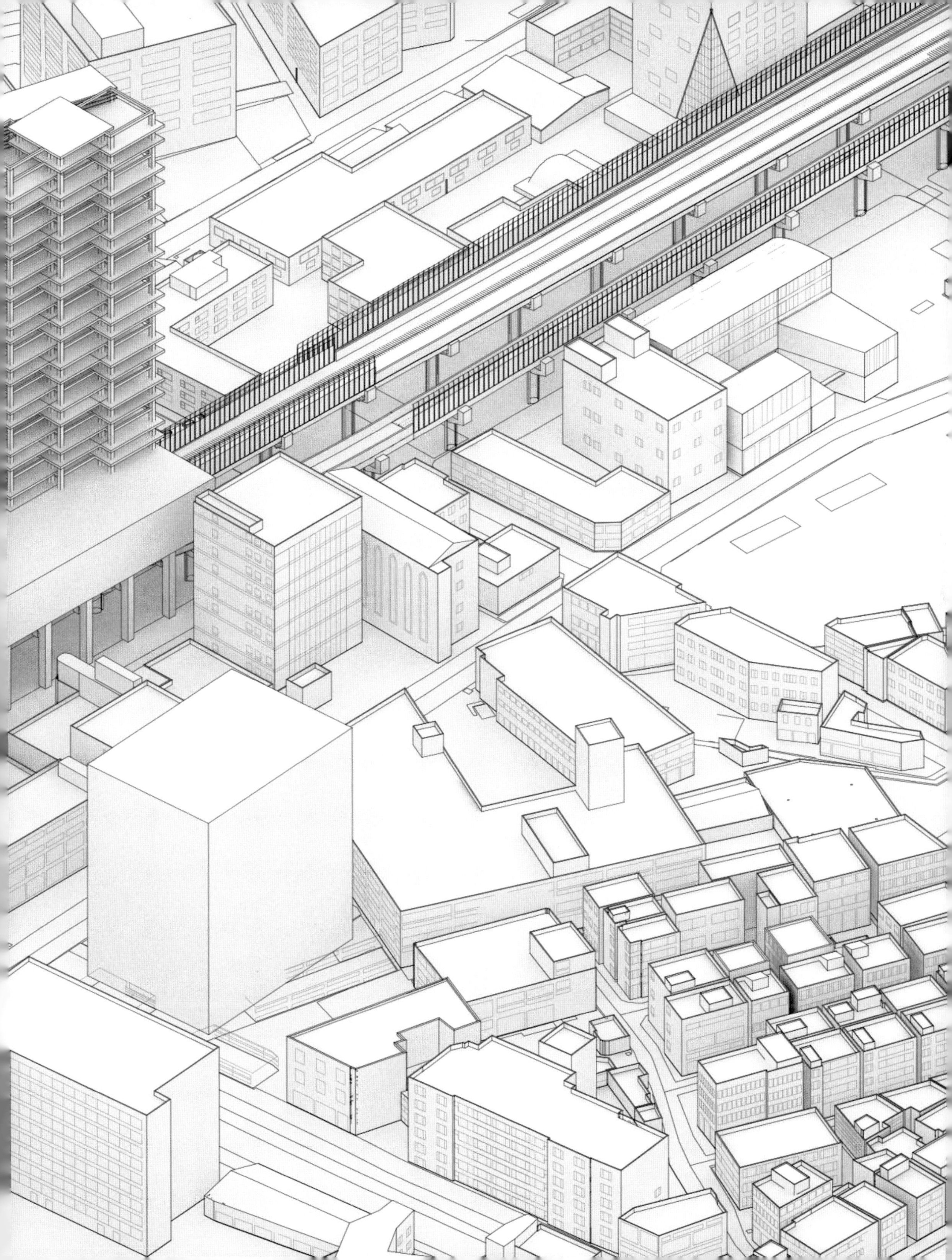

SUBWAY DISTRIBUTION HUB IN SEONGSU

Namhyuk Baek

This project attempts to solve the current decline of Seongsu's industry by setting up a system of transporting goods that interfaces with the human transport system. The project defines so-called good design as deceiving and consuming, and focuses on the essential relationship of real-estate and physical networking.

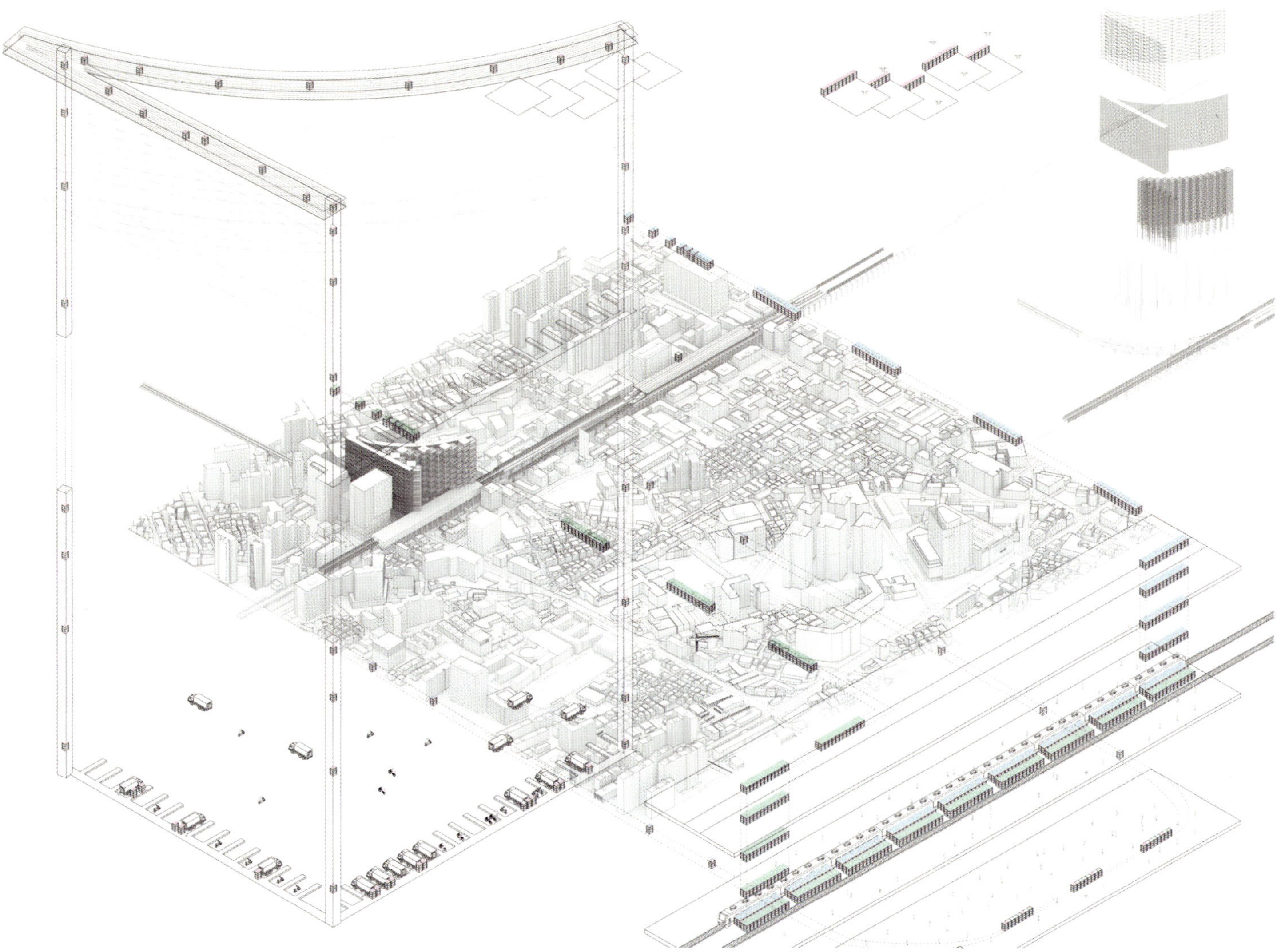

1

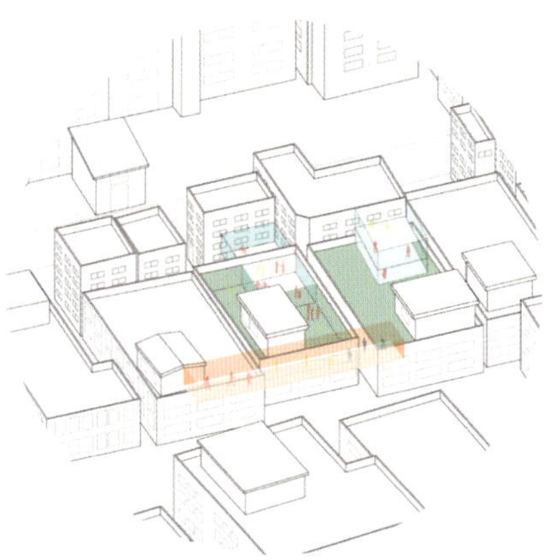

2

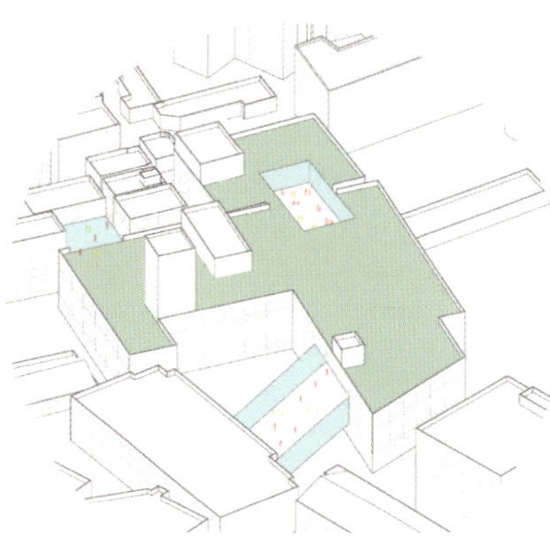

3

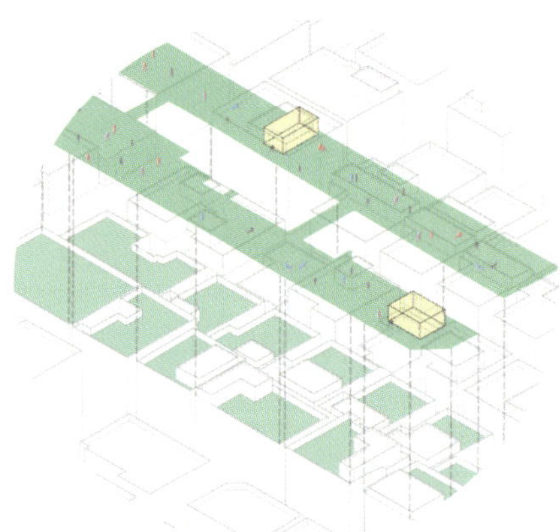

4

1. System and Structure Masterplan

2-4. Possible networking space
The current distribution system which uses cheaper lots outside the city center results in the loss of efficiency, and Seongsu's infrastructure should be delegated for the such : by reusing its subway infrastructure and the currently grey space above, it will become the main hub for the new subway distribution system.

5-6. Connection : Warehouse - Units

7. Connection : Units - Units

8-10. Plans (Factory/Residential/Office)

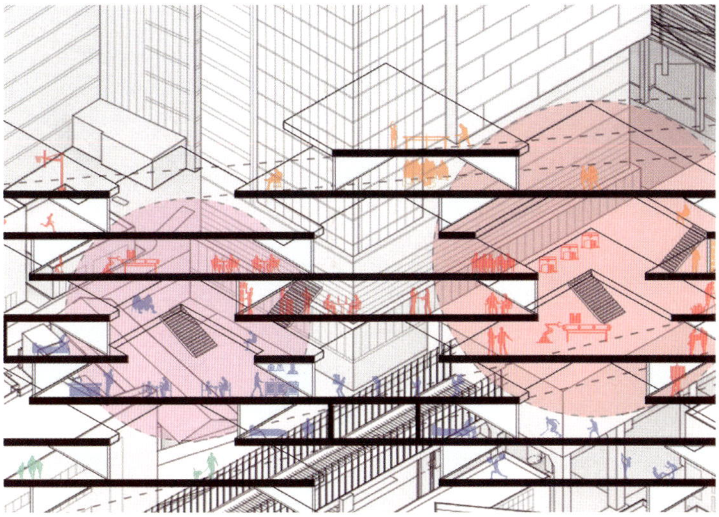

5

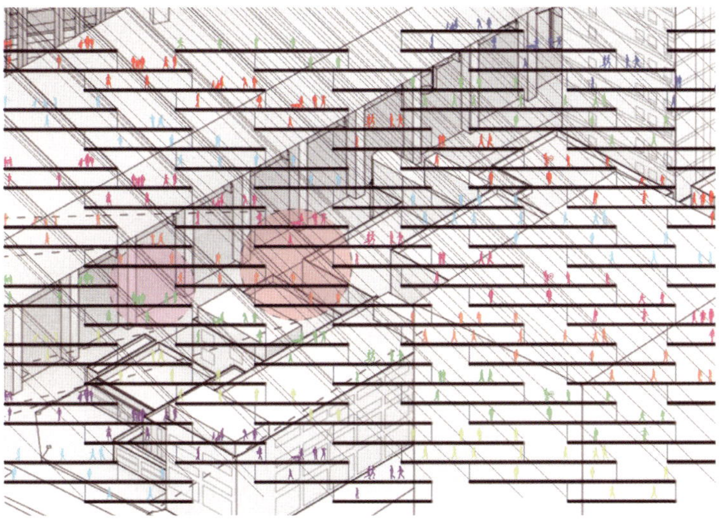

6

7

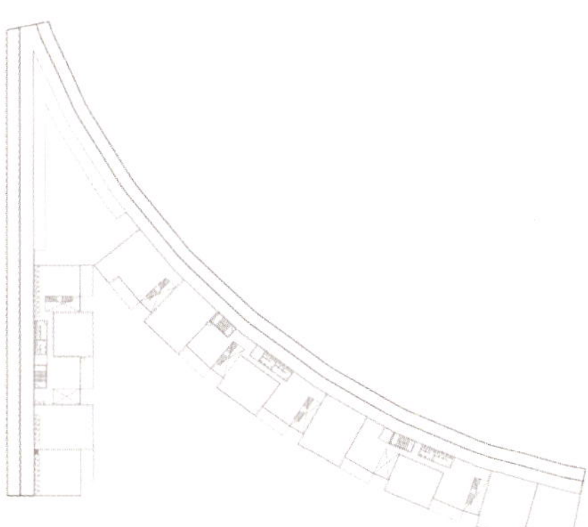

8

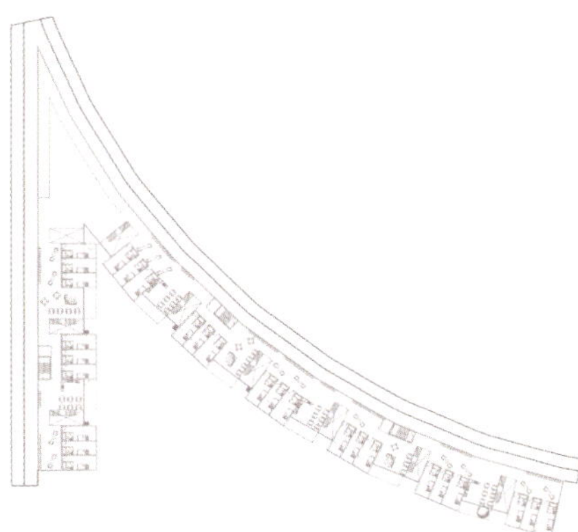

9

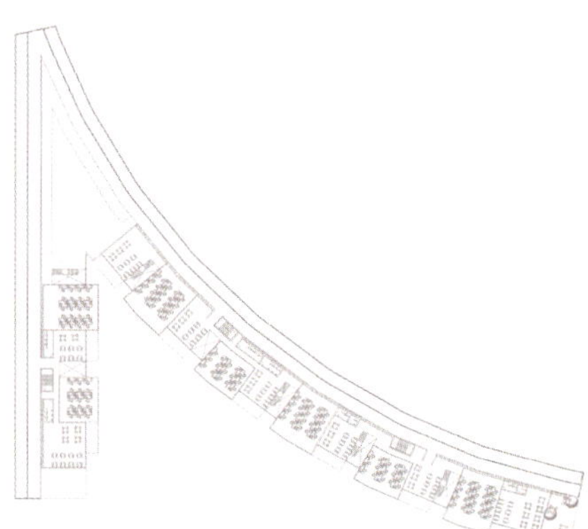

10

11. Unit Arrangement
The hub is mainly a warehouse but its program greatly differs from the typical : it also contains residence, office, and even manufacturing units. The public space for each program owes its character to what item it receives from the distribution system, which increases chances for interaction, as well as efficiency in transportation of the products.

12-13. Model shot

11

12

13

14. Model shot
Seongsu, which once had been a main supplier for the whole domestic manufacturing market long ago, will revisit its capacity to literally provide for everything in need, through not only production but also transportation.

14

ROOFTOP NEIGHBORHOOD

Keonhee Kim

This project understands the transformation in industry in Seongsu, and discusses a neighborhood for the such. Seongsu, designated as a semi-industrial area rarely found in nowadays Seoul, still bears plants fabricating metal, as well as its hand-made shoe industry. But a change including emergence of cultural spaces is happening, along with the decline of its original industry. Furthermore, Seongsu faces segregation among programs, accelerated by its raised infrastructure which not only dissects the area visually, but also in terms of walking.

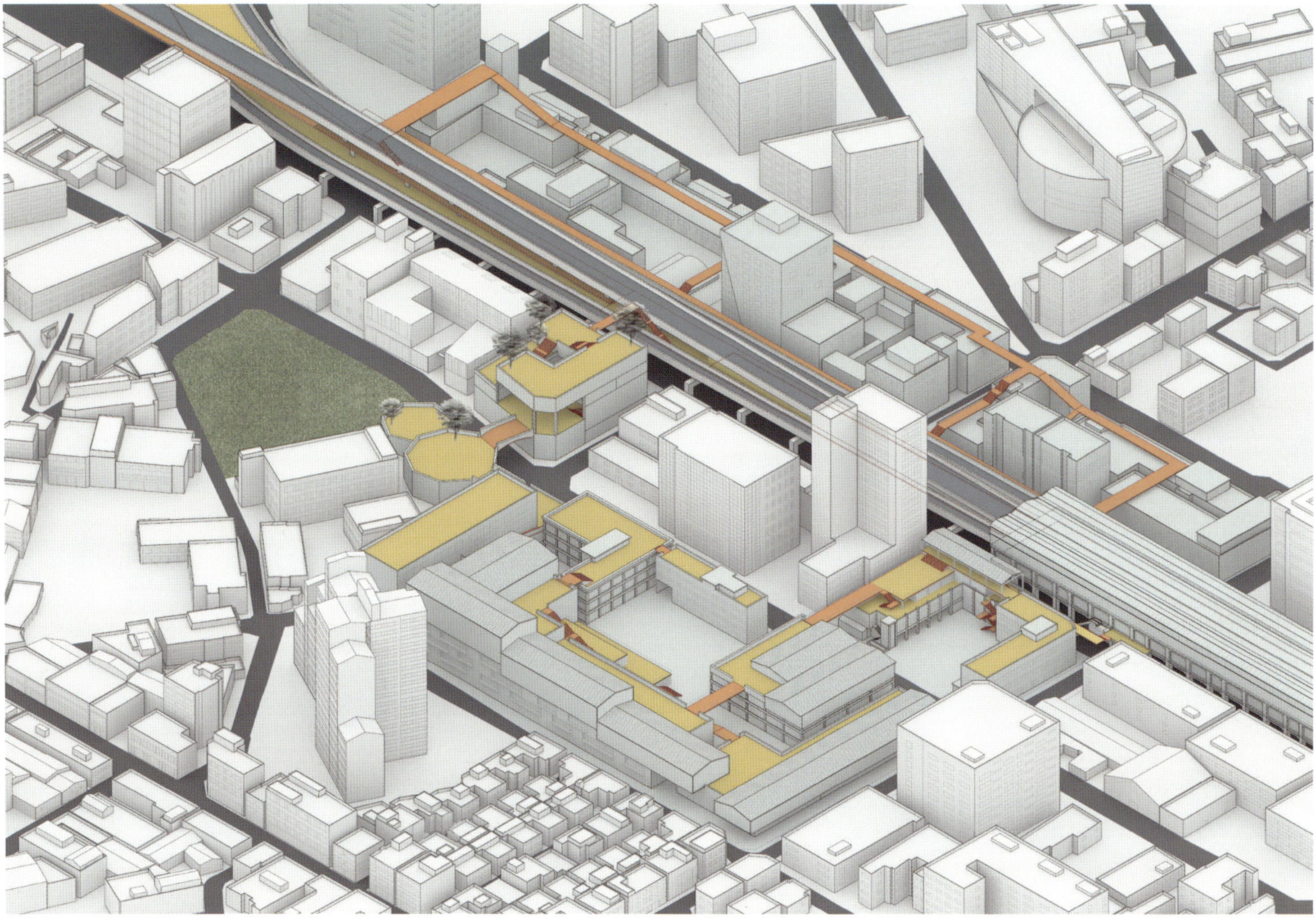

1

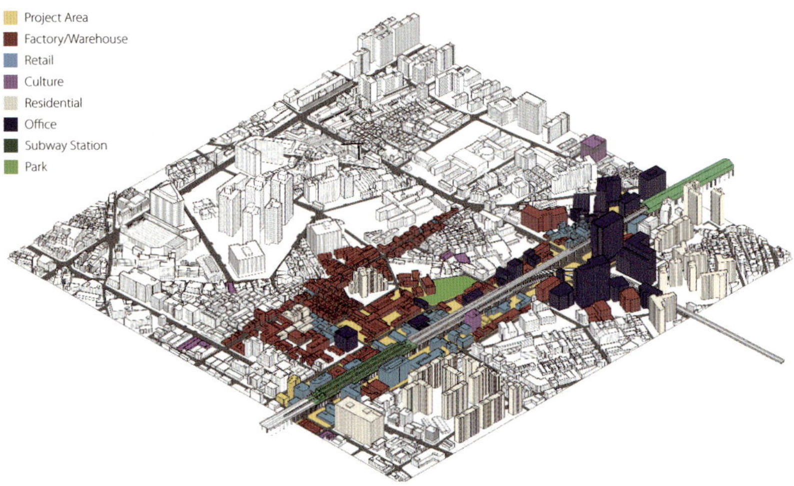

2

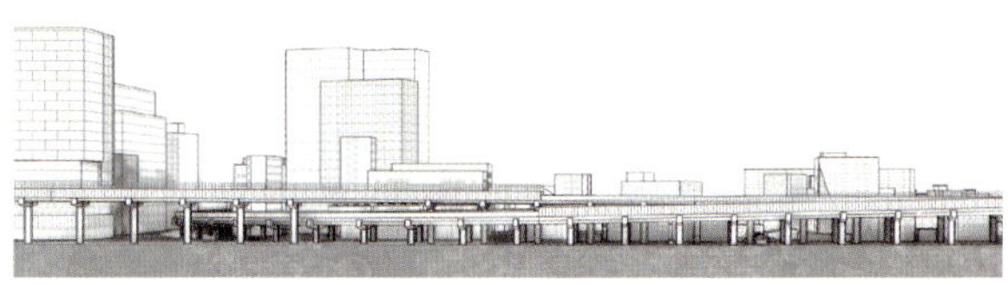

3

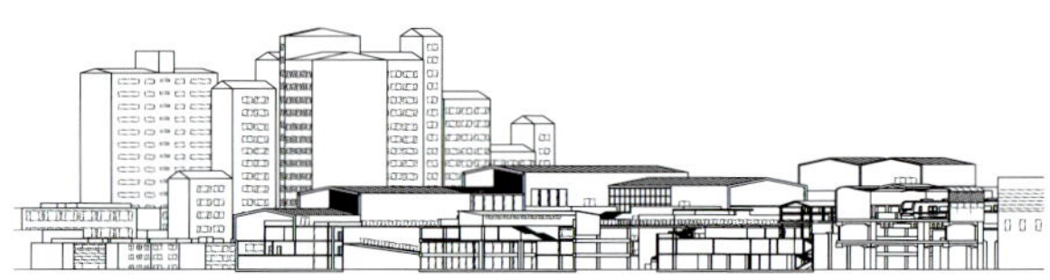

4

1-2. Masterplan

3-4. Sections

5-9. Perspectives & Plans

The project transforms the grey space under the infrastructure into a walkway and assigns a new community by connecting the rooftops of existing fabric. The public will access buildings through both the ground and the rooftop : while the former consists of commercial space, the latter promotes acquaintance among providers, designers and craftspeople. Improving such relationships will provide the manufacturing industry - which has been reluctantly taking on 1-directional sale system - with a decent working system inside which subdivided labors can efficiently cooperate. Residential and educational programs supporting start-ups are also provided, all of which add up to the new neighborhood concept.

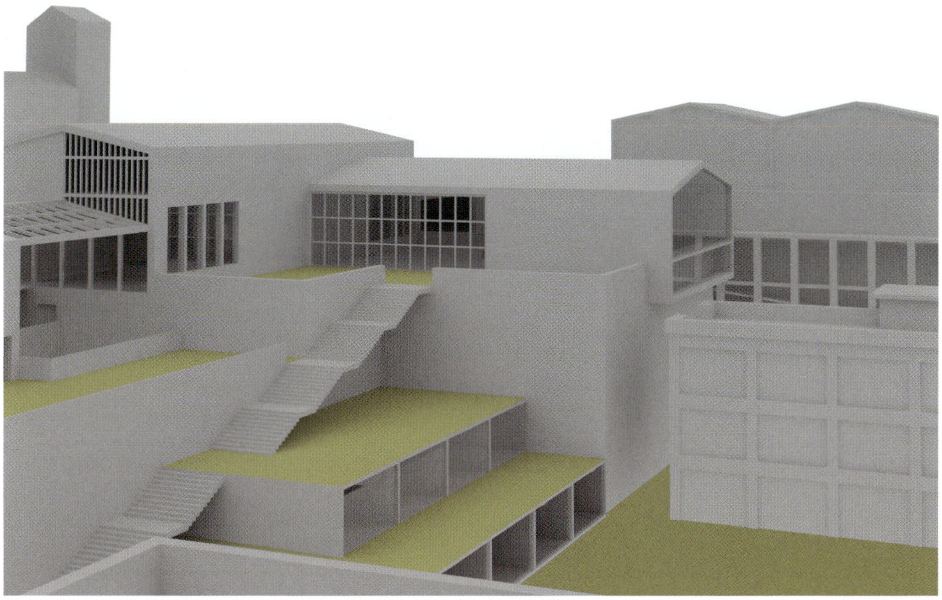

5

6

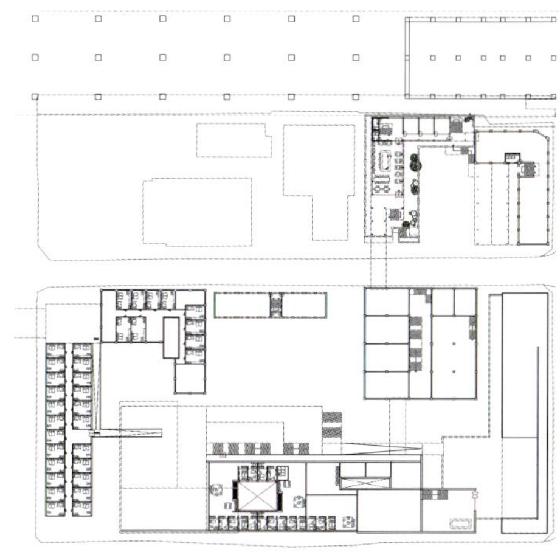

7

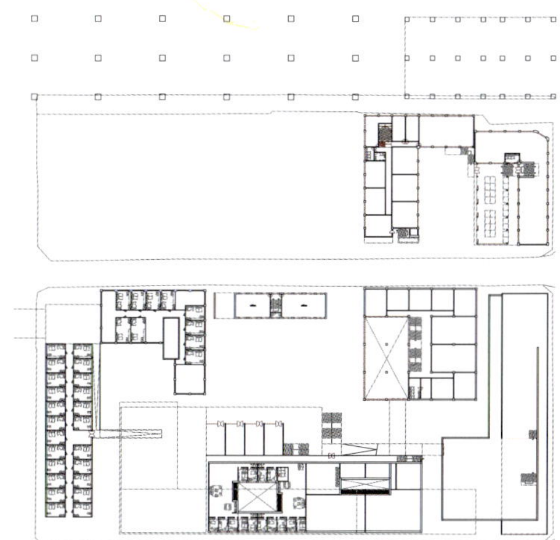

8

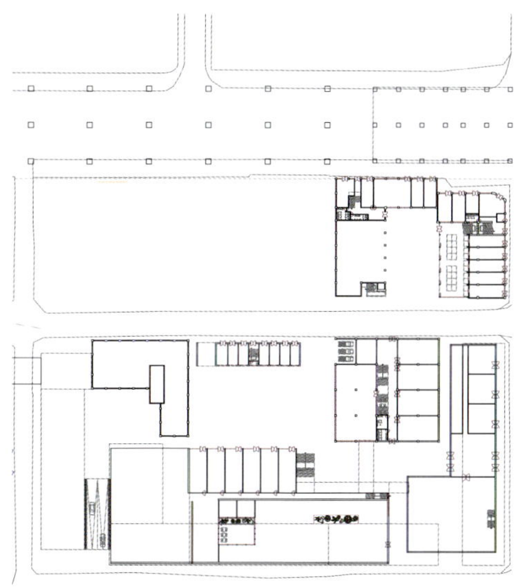

9

URBAN SYNECDOCHE

Youngjoon Shim

This project identifies Seongsu's history as more autonomous than mandated, with its urban fabric having been constantly customized and fragmented according to its people's need. But the decline of the hand-made shoe industry and the tendency to relocate industrial areas out of the city center has resulted in its shrinking, and this project aims to revitalize each block and labor sector by urban customization.

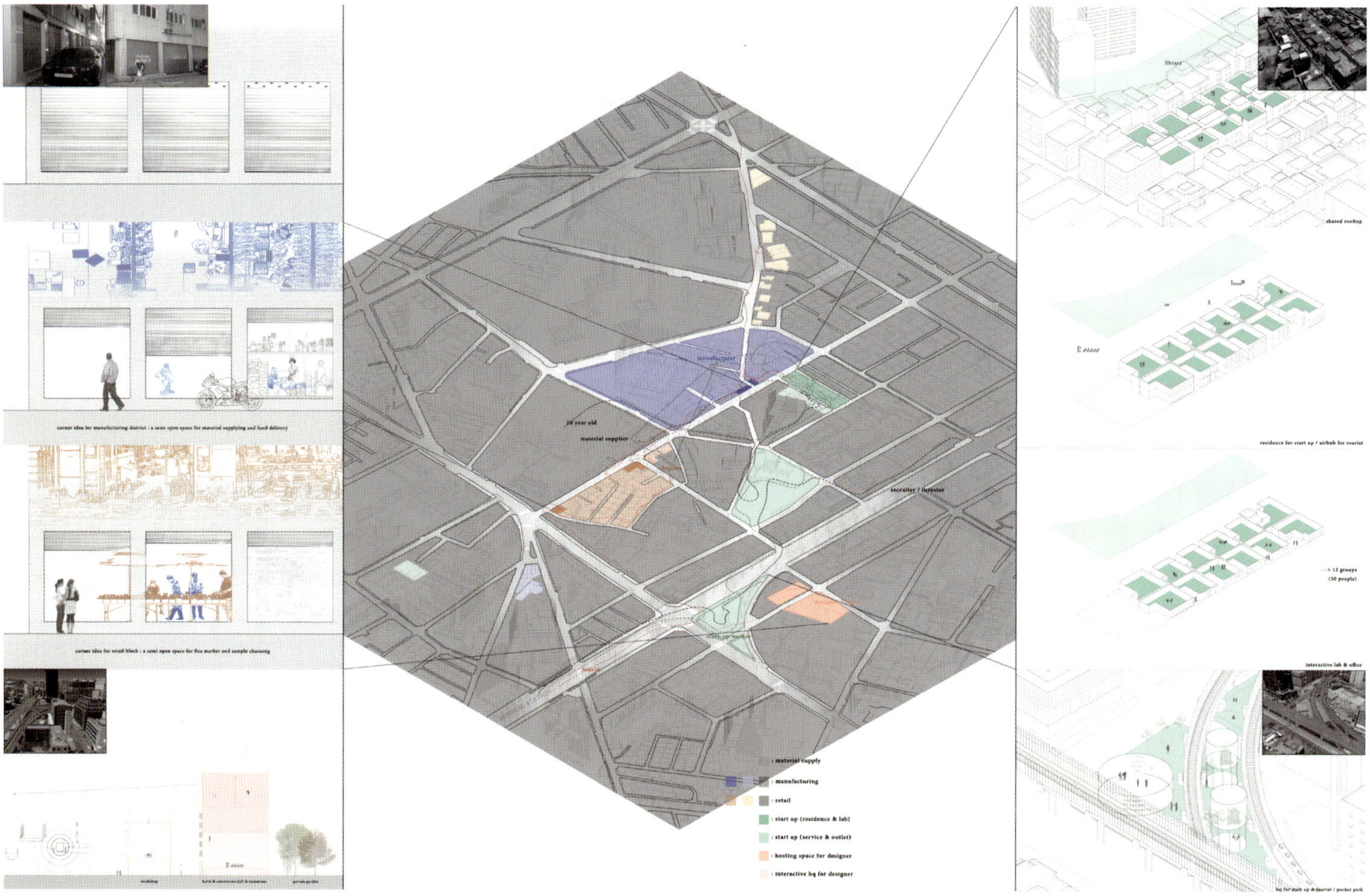

1

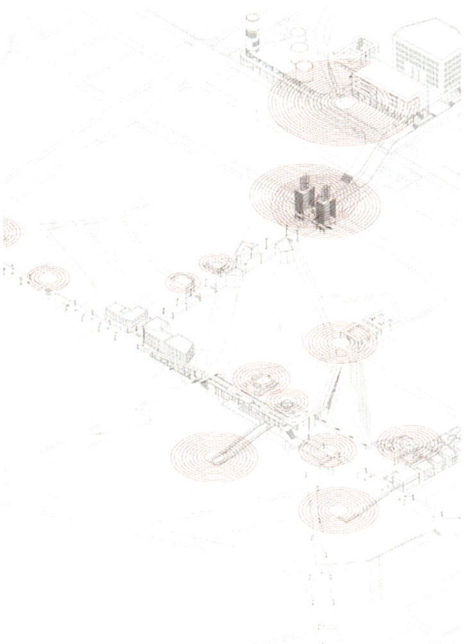

2

3

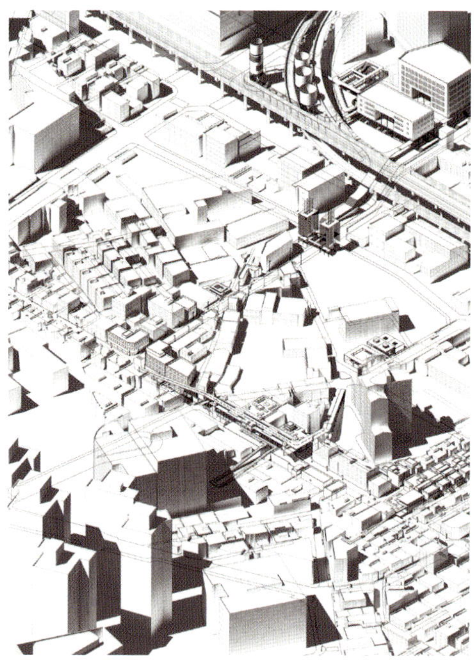

4

1. Masterplan

2-4. Layers of Synecdoche

5-6. Final Synecdoche
The systematic approach to accomplish such goals resorts to the notion of synecdoche, a figure of speech where a part of something stands for the whole. The project understands the city not as a concrete and tangible object, but as an interpretation deduced from each person's experience and memory, while physical intervention is applied to the degree to which such understanding and deduction is made possible.

7-8. Street transformation over time

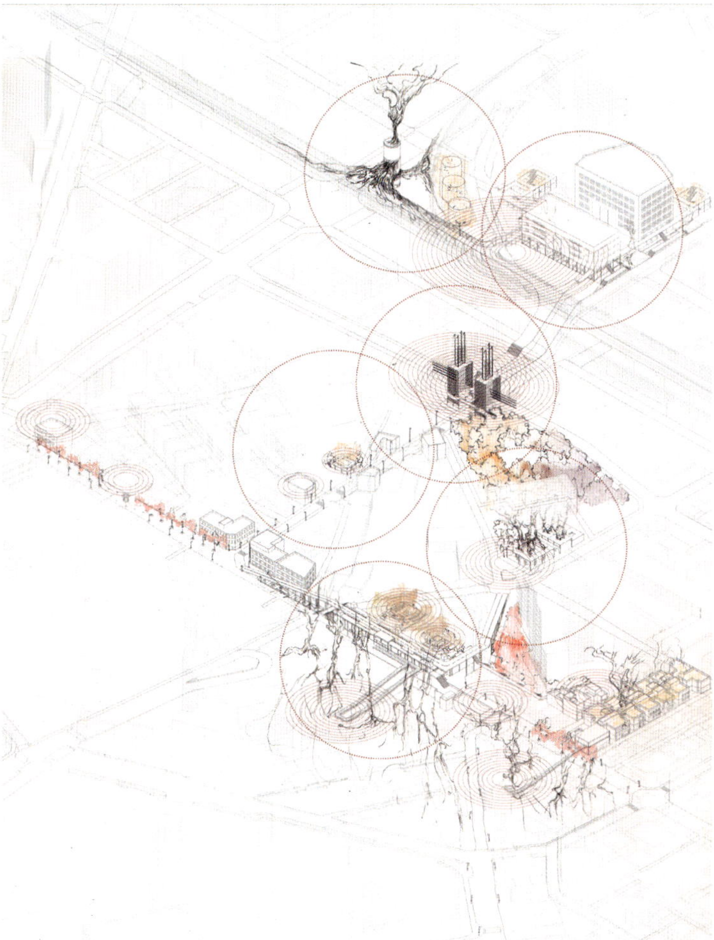

5

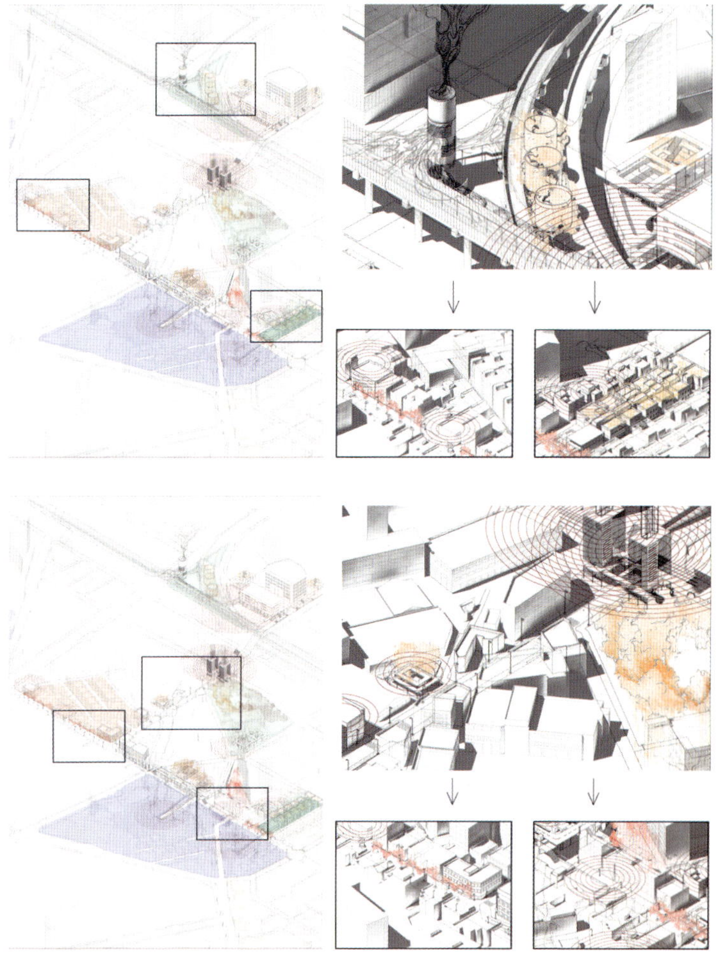

6

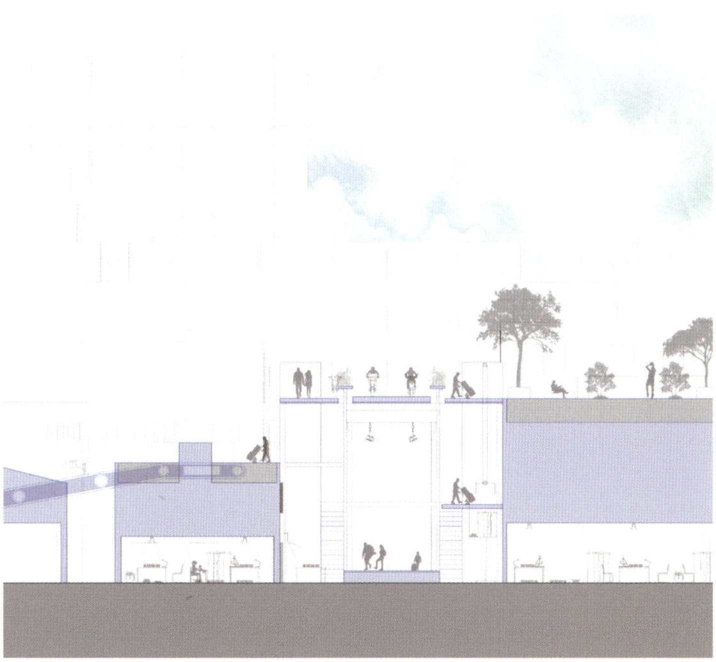

7. Manufacturing Street - 3pm

8. Manufacturing Street - 8pm

9-10. Plan in Detail - Residence for Start-ups
That is to assign each labor into each block, so that each subdivided labor, including new incomers such as those in the start-up and fashion industry, can customize and develop their own block accordingly.

11. Dispatching Overpass and its components
The materials dispatched into the site will be digested in many possible ways, and foreigners, as members of the new 10-min neighborhood concept, will operate the district efficiently, in order to attain their goals, while acknowledging and contributing to the future industry of Seongsu.

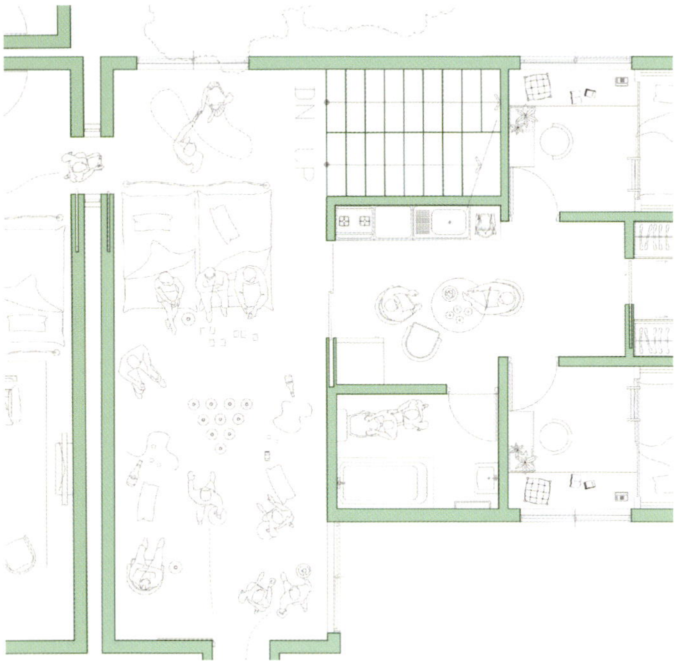

9

10

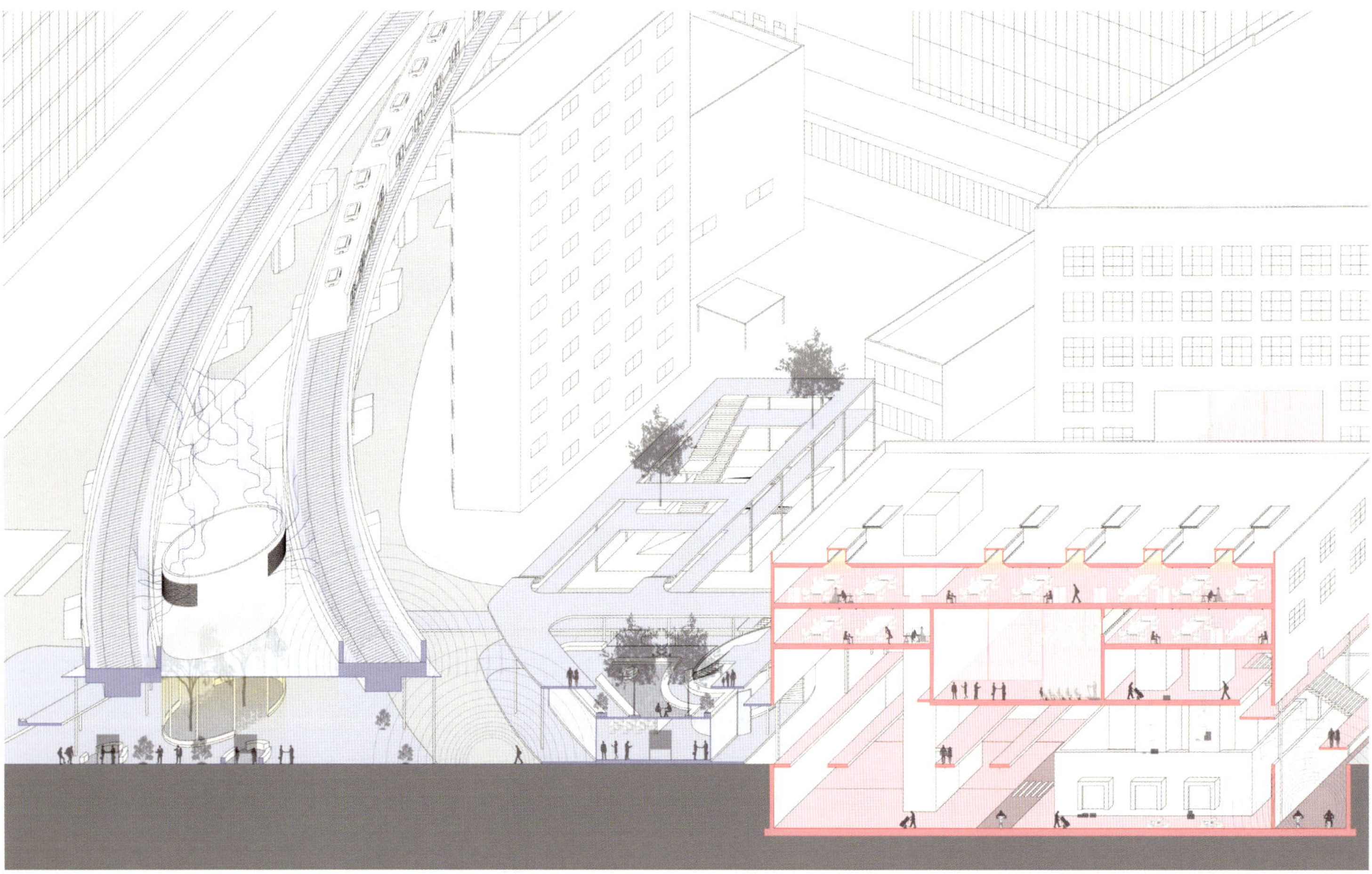

11

12. Perspective - Interactive Overpass

13. Perspective - Dispatching Overpass

14. Perspective - Reflective Vista

15-18. Projection Model
Partial interventions including overpasses, pavilions and even street lights that create visual, sonic, and aerial phenomena, may or may not stand for the concept of the whole city, where each citizen becomes a perceptible interpreter.

12

13

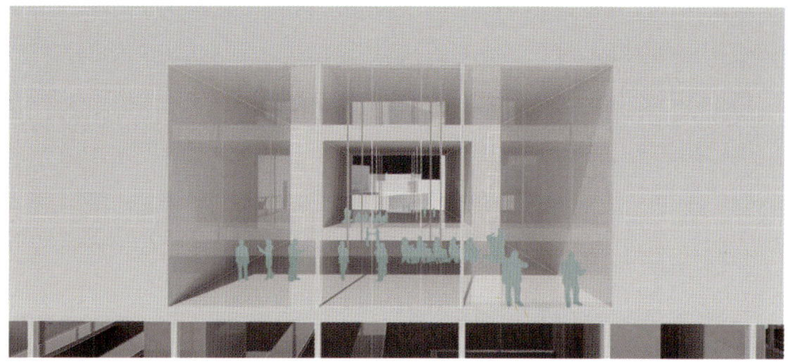

14

15

16

17

18

SEOUL CYCLES

Trygve Schmidt Pedersen

This project aims to integrate multi modal transportation into the everyday life of Seoul, in line with its own interpretation of 10-min neighborhood concept focused on interactive space and human scale. It acknowledges the cyclist's accessibility, and provides for a broad range of cycle paths : Seoul Cycles, reusing the lost space above, along, and underneath the existing subway infrastructure, contributes in making Seongsu a united neighborhood, along with enhancing cyclists' safety.

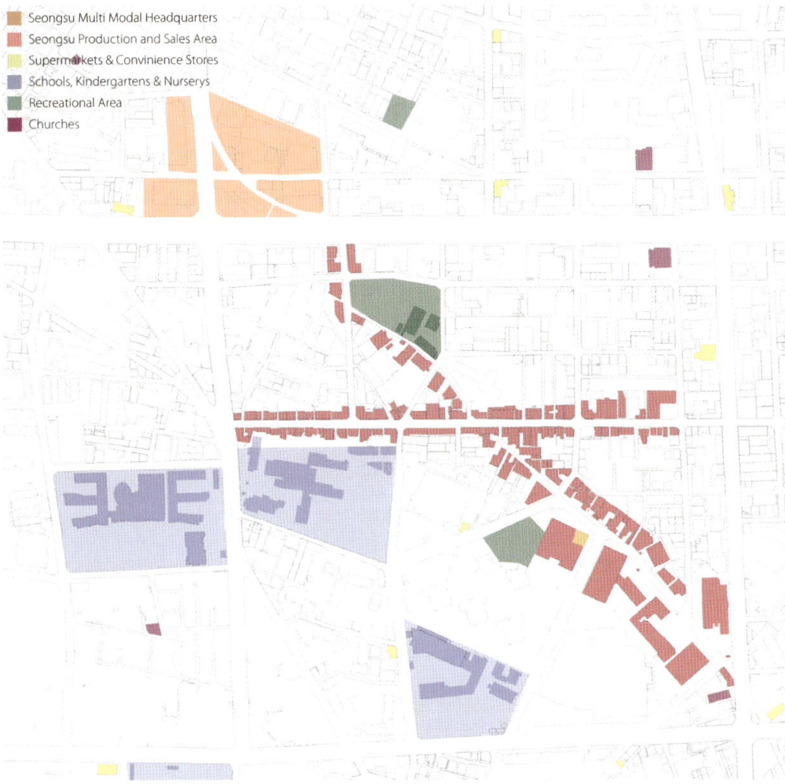

1

2. Existing Cycle Grid

3. New Cycle Grid

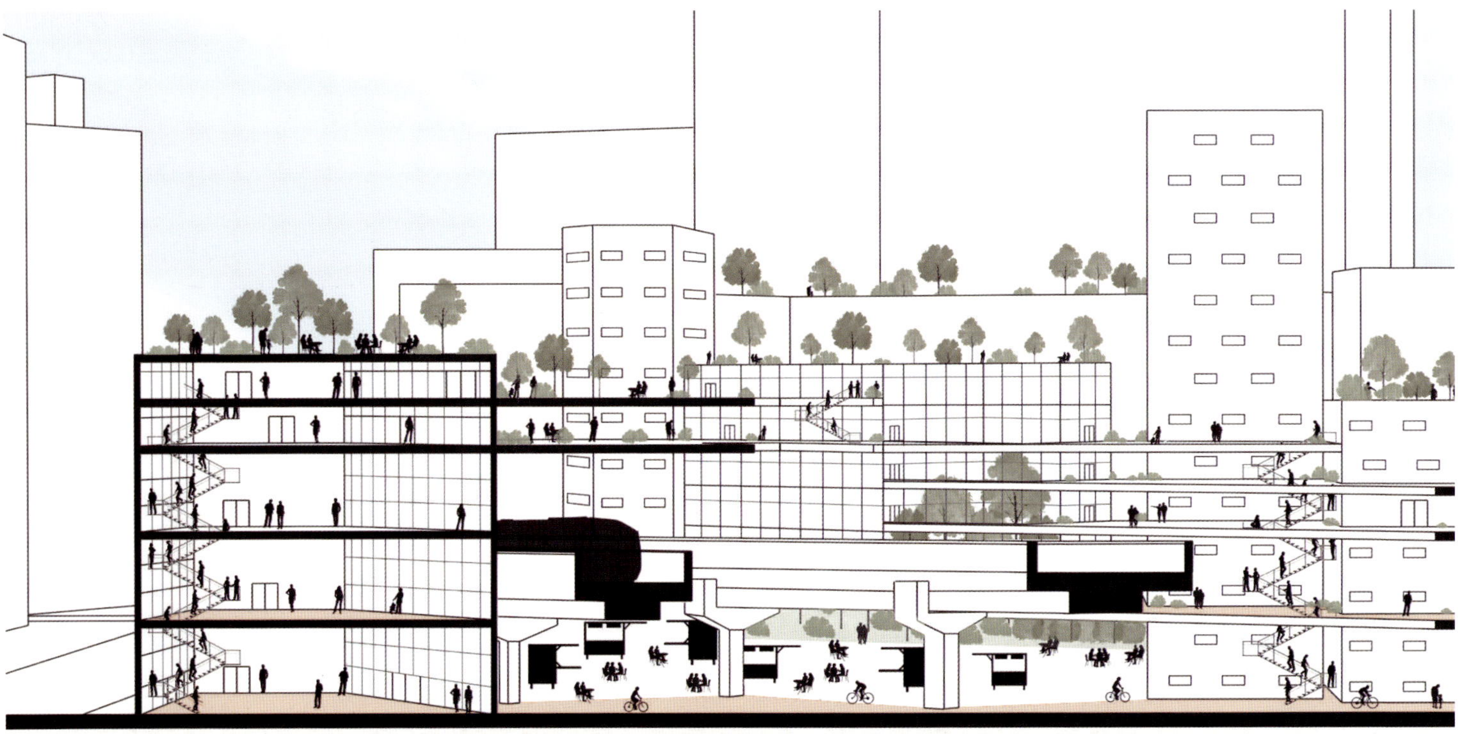

4

1-3. Masterplan

4. Section Perspective

5-10. Plans & Model shot

The site is also provided with a Multi Modal Headquarters mainly promoting connection between every means of transportation, which increases not only mobility but also invites many and various types of people into the site as well. Accordingly, the headquarters bears communal spaces including library, jjimjilbang, office for start-ups, and even kindergarten, opening up the program to not only local residents but also visitors via convenient multi-modal access. This again emphasizes the idea of the neighborhood unit, where the attempt to recover the sense of interaction in human scale, through every appropriate means, is prominent.

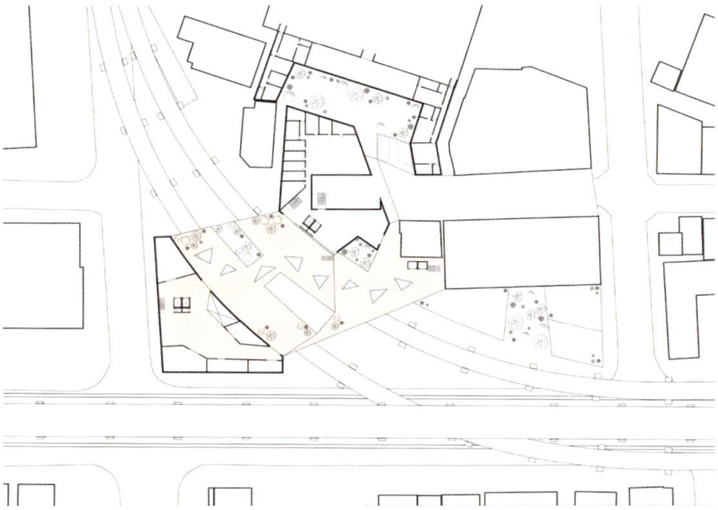

5. Top Floor

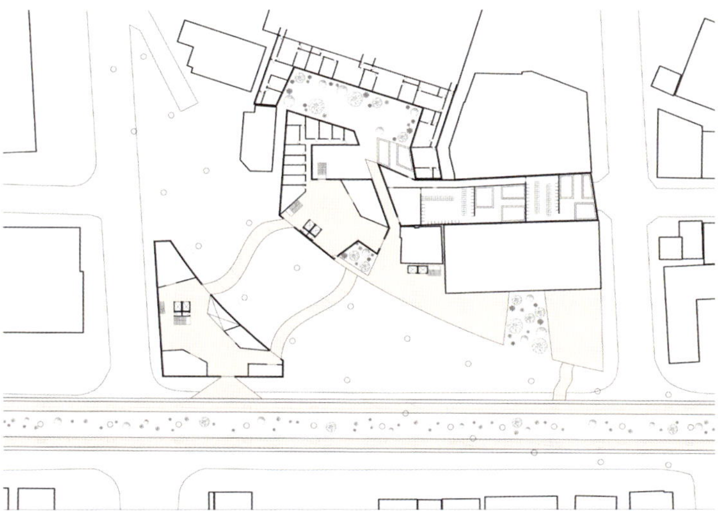

6. Transportation Floor

7. Ground Floor

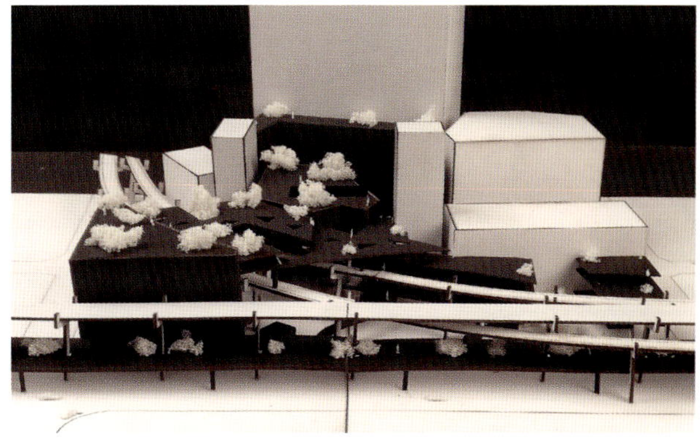

8

9

10

FIELD OF WALLS

FIELD OF WALLS

Daye Kim

This project focuses on Seongsu's regional characteristic as a semi-industrial area located in the city center, and the fact that knowledge-based industrial center as well as apartment factories are built regarding this characteristic. The semi-industrial area aims to realize coexistence of industry and other functions, while resolving conflicts arising from the mix.

1

2. Riverside Park

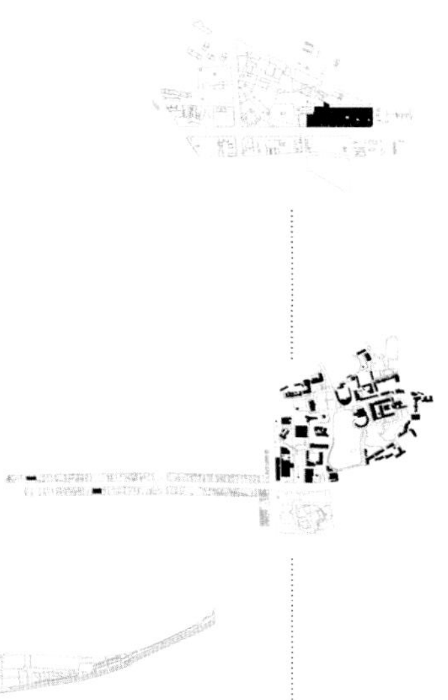

3. University

4. Seoul Forest

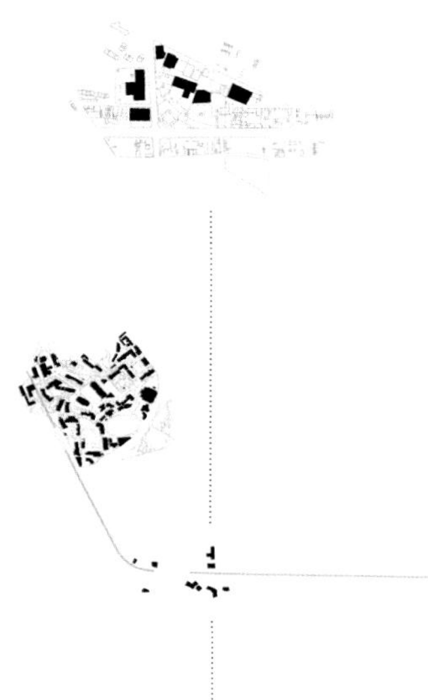

5. University

1. Flow of Infrastructure and People
In this regard, the project is based on the conviction that the knowledge-based industry center is able to play a major role. To be specific, support facilities will supplement functions injected into the neighborhood including education, commerce and culture, along with the concentration of various industries.

2-5. Related Urban Forms

6. Isolation of Lots - Landscaping and Orientation
The project looks to the ground level of the knowledge-based industry centers, which until now has been aimed at maximizing individual profit, as having potential when reused as a public space.

7. Isolation of Spaces - Car Circulation

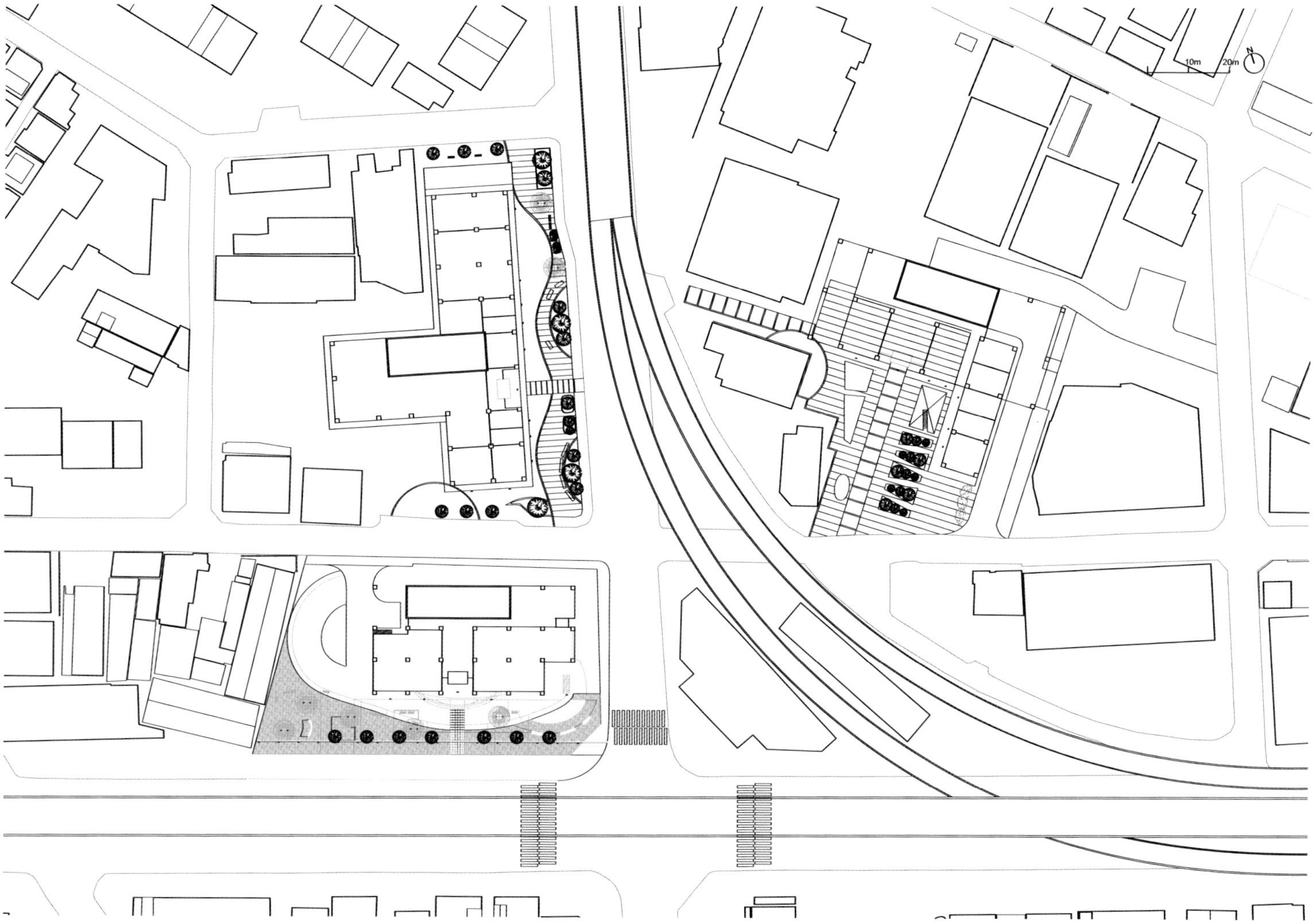

6

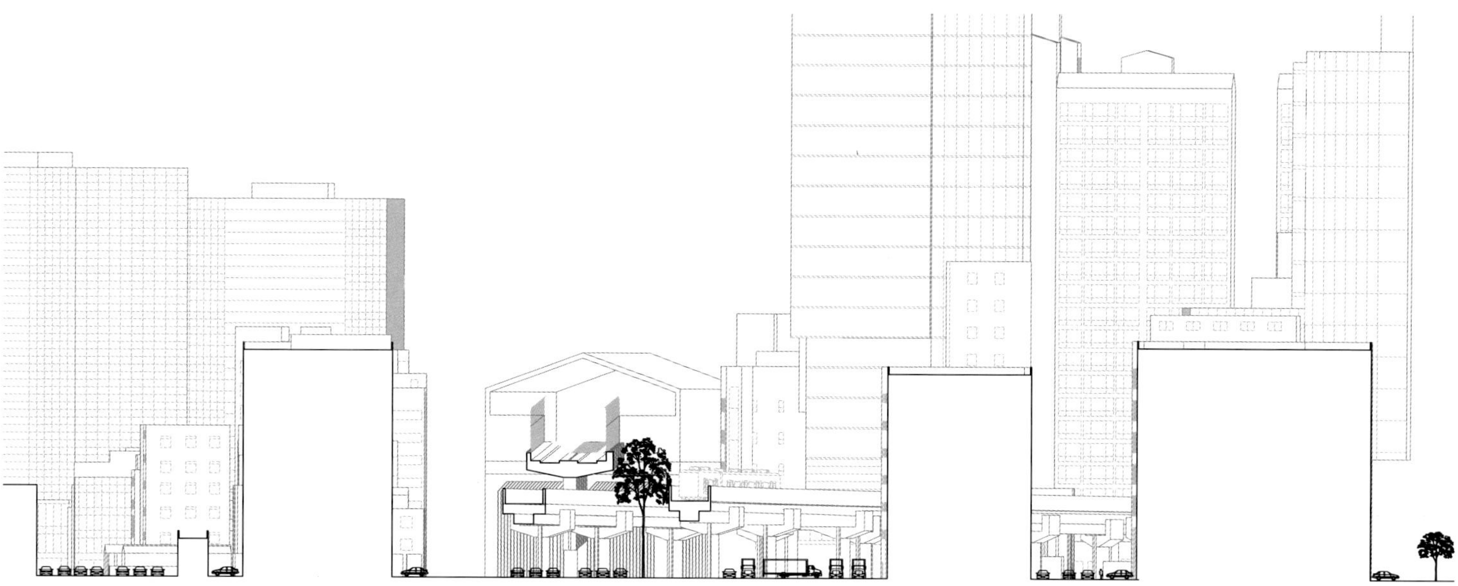

7

8. Current Urban Fabric in Detail

9. Re-imagining the Urban Fabric

10. Possible Circulation with New Interpretation

11. Pedestrian and Car Circulation
By introducing aboveground and underground vehicular circulation, parking strips and lots are removed from the ground level.

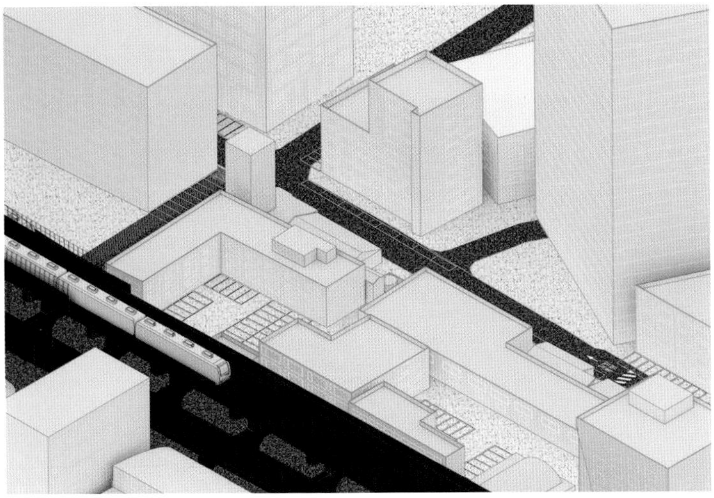

8

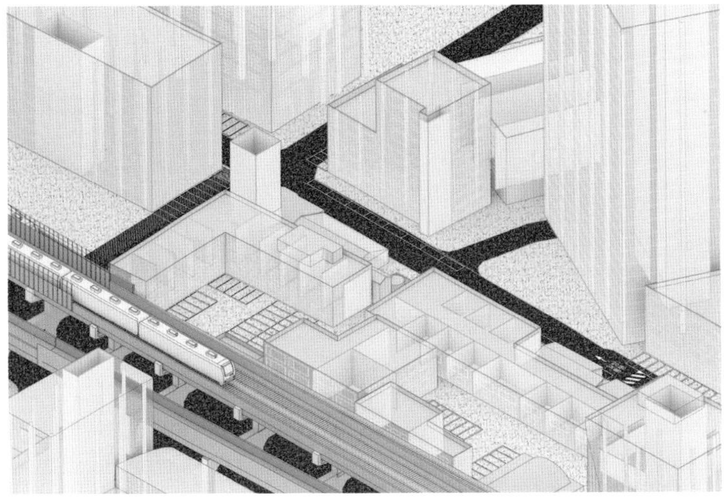

9

10

12. Possible Circulation through Intervention
Next, the bottom part of the existing center is voided in order to promote network density on the ground level, while currently underused lobby and neighborhood facilities are given a floor pattern that allows the existing structures to be newly understood, both individually and as a whole as part of this expanded horizontal territory.

13. Ground Floor Plan
As a result, a newly accessible ground level is introduced inside existing buildings, allowing start-ups and commercial facilities to be exposed visually, while pedestrians are invited into these open spaces to view the distinctive cityscape of Seongsu.

11

12

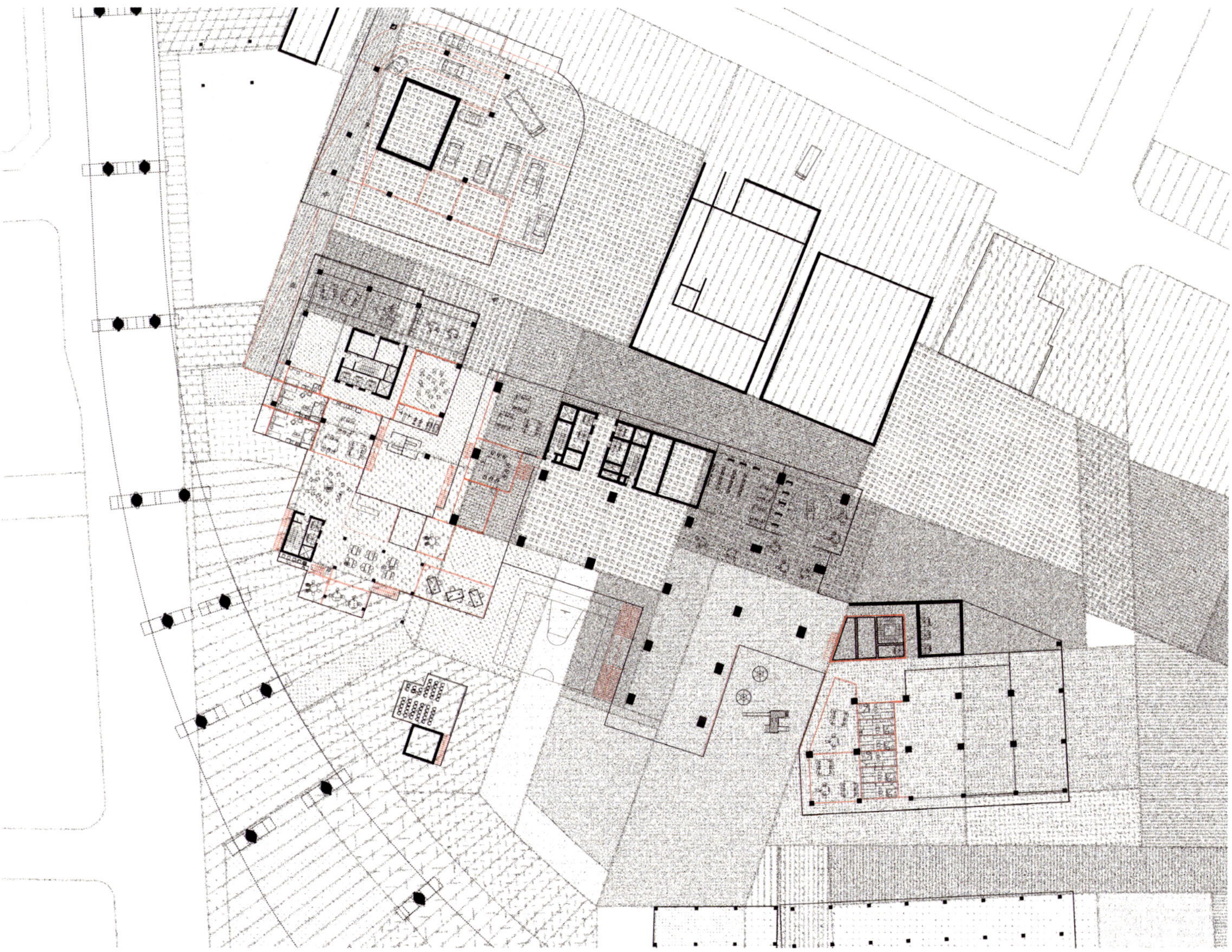

13

URBAN STITCHING

Soobin Kim

Facing inevitable decline due to the growing popularity of foreign brands and mass-production, Seongsu is re-imagined to support production units for wearable technology and promote the free exchange of ideas through a creative re-knitting of the city on the urban and architectural scales. Programmatic requirements in doing so are identified using the voxel, which is then applied to various existing buildings within Seongsu.

1

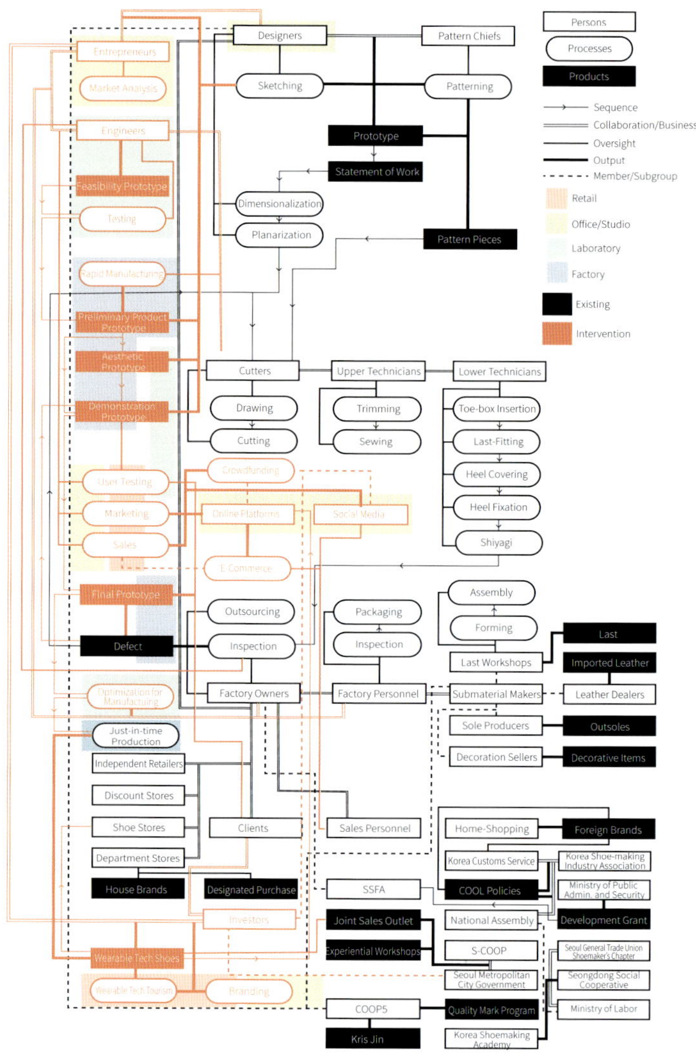

2

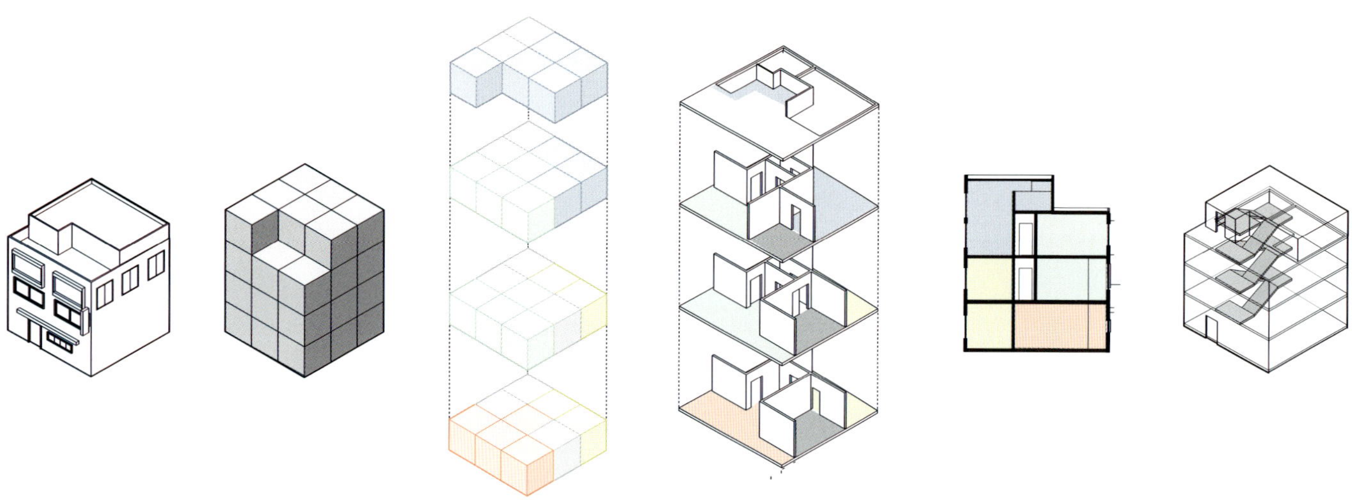

3

1. City in Voxels
Voxels allow a much more fine-grained designation of use than zoning and brings about truly mixed-use buildings without compromising livability.

2. Shoe making Ecology Map
A technology-based intervention strategy was added to the existing shoe making ecology to identify the necessary spatial typologies.

3. Voxelation Prototype

4. Adjacency Guidelines
Certain adjacencies are recommended to improve working conditions and foster networks among different specializations and generations.

5. Voxelation Prototype Expanded
Voxels for housing, shared amenities, and circulation have been added.

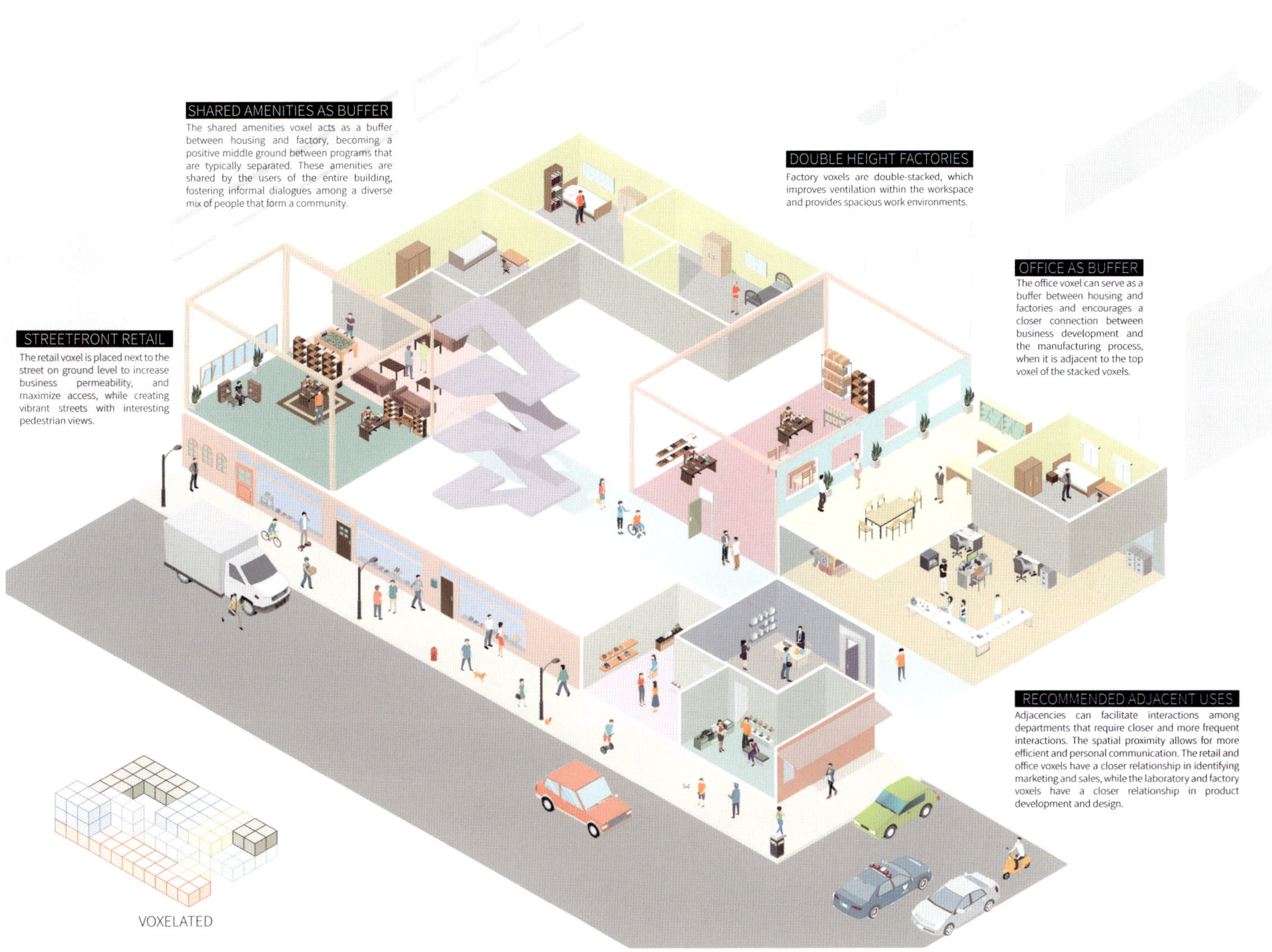

4

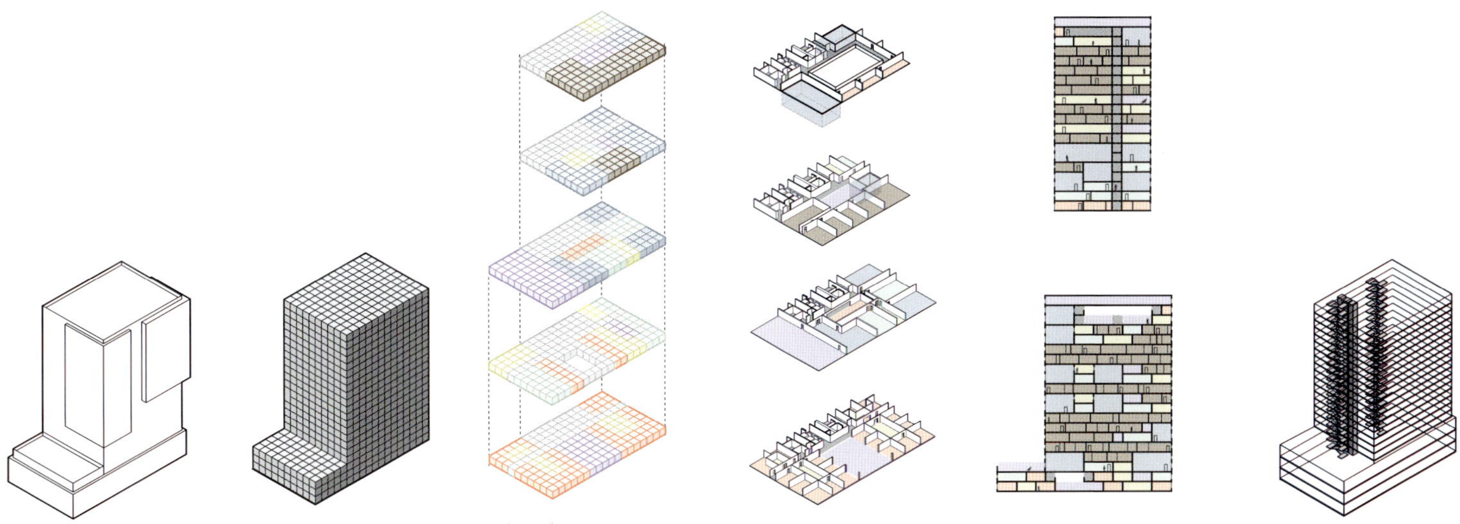

5

6-9. Perspectives
The under-bridge site in Seongsu-dong is developed into a learning center that sits on top of an urban park. The ground level is lowered to cultivate a lush urban forest, where small classrooms peak out in between the foliage. The auditoria seating gives form to the building, while a unique circulation system enwraps the structure to provide continuous access to the park.

10. Building Section

6

7

8

9

10

11-13. Plans

14. Ground Floor Plan

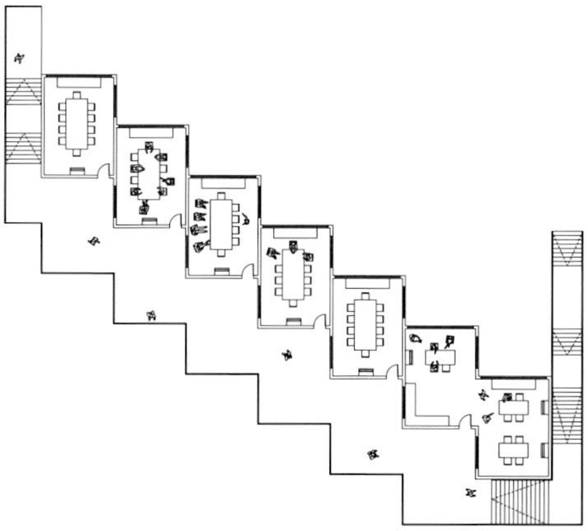

11. Classrooms

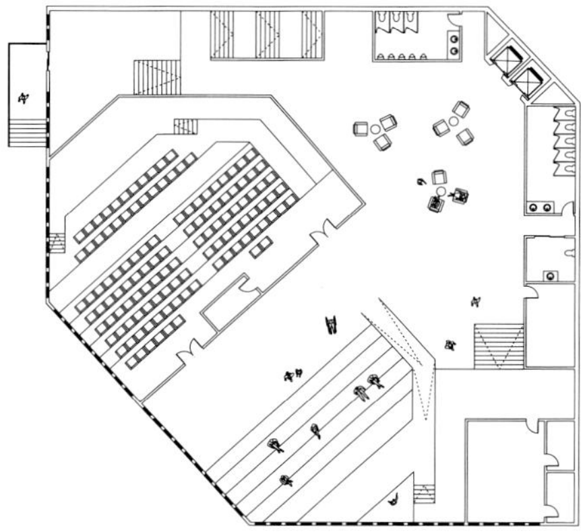

12. Auditoria

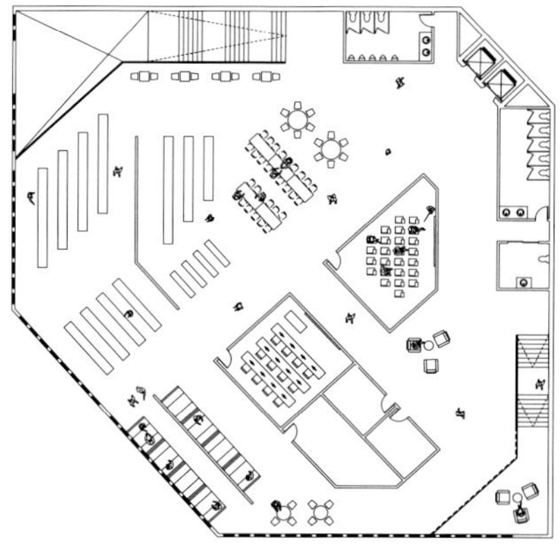

13. Learning Center

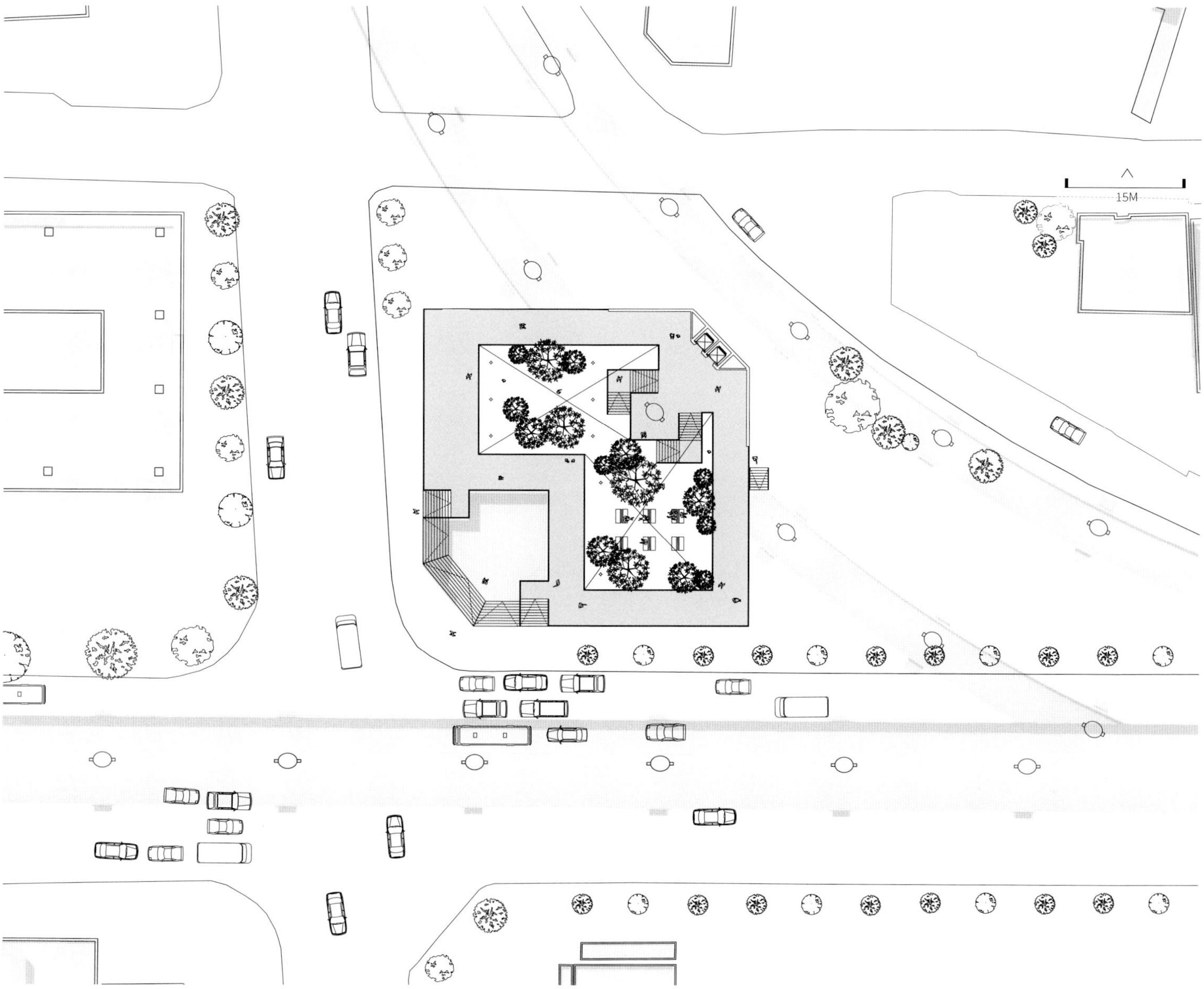

14

3 DIMENSIONAL DIVERSITY

Bonghoon Jeon

This project explores the idea of the vertical city, against the existing urban fabric which is unnecessarily horizontal, thus confronting the problem of urban sprawl and low density on the city's boundary. Even the city center, which may be exempt from such problem, still suffers from lack of diversity in section, resulting in high-rise buildings and infrastructures disconnected from everyday life.

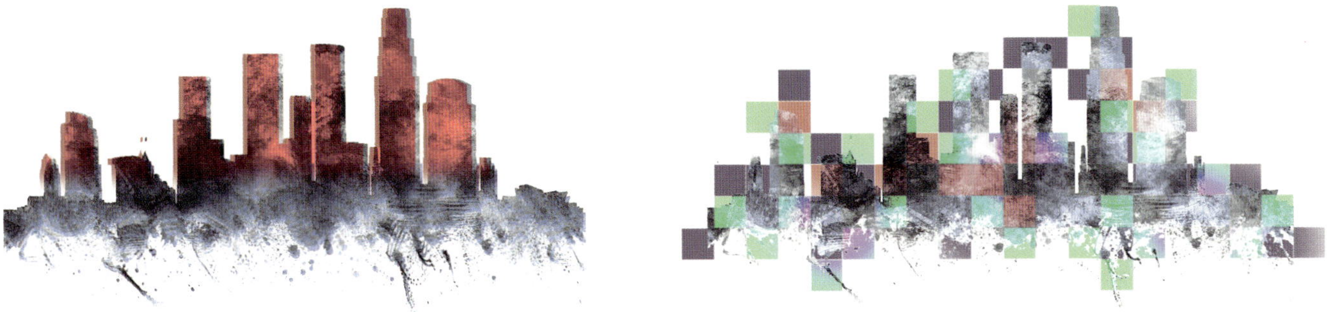

1

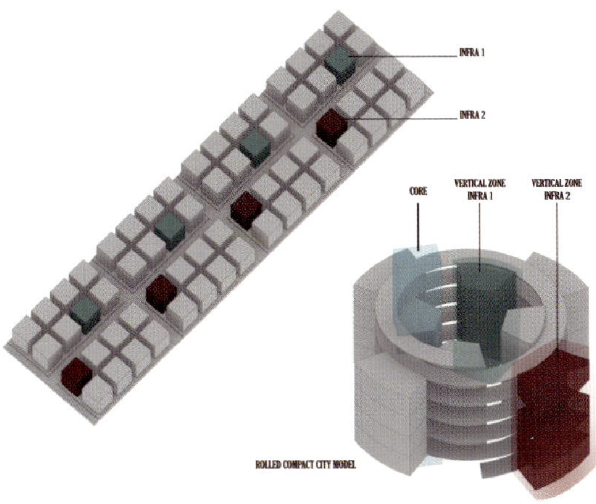

2

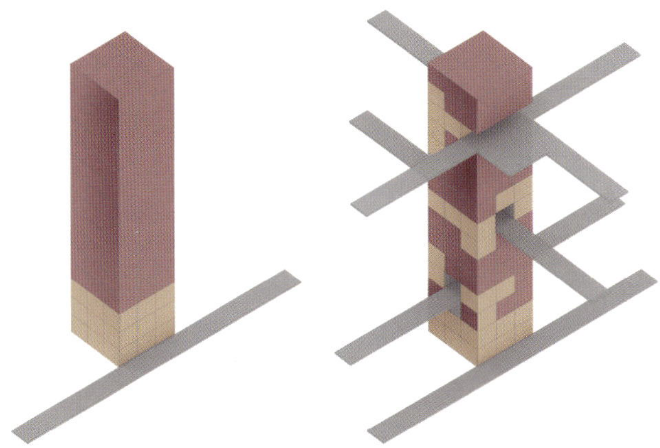

3

4

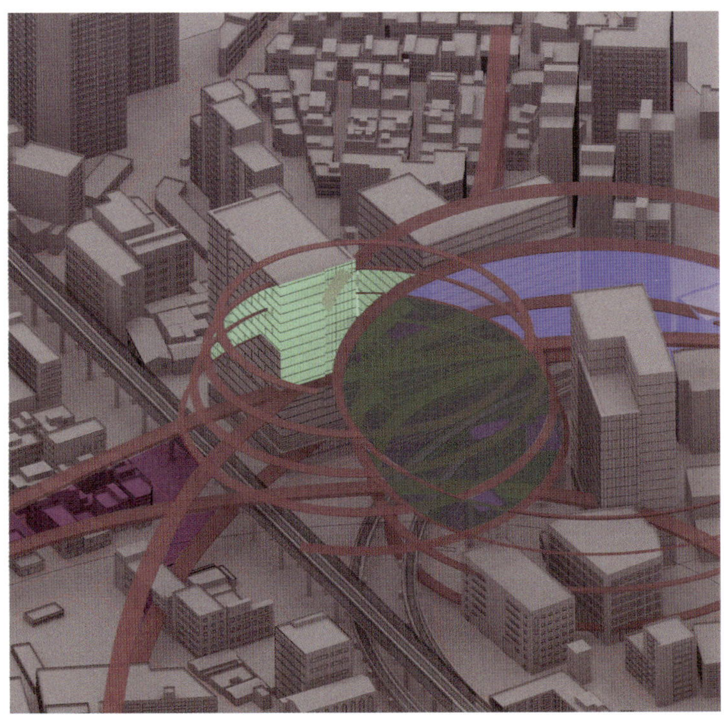

5

1. Cityscape

2-5. Related Programs & Schematic Diagram
The vertical city – aimed at recovering urban diversity by introducing the vertical axis – uses the spiral as its main language. Rolling up the traditional linear urban fabric, which inverts the idea of the 2 dimensional walking plane, allows for more efficiency and diversity.

6-8. Section & Perspectives
Infrastructure, branching out to even the upper levels of the high-rise buildings, will invoke erosion of program around the penetration, overcoming the mundane public-vs-private distribution along the vertical axis. Seongsu will become a dense city comprising of urban spirals, each of which bears its distinctive character and the intersection of which bears a mixture of programs, all adding up to diversity in 3 dimensions.

6

7

8

TRANSITION

Ina Choi

This project zooms into Seongsu's various lives, where invisible conflicts among residents, workers and cafeteria owners pervade, and a need for a better living and working condition is present. The project unravels the situation by looking at the potential of the underused overpass area and the historical underground stream that once existed prior to its construction.

1

2

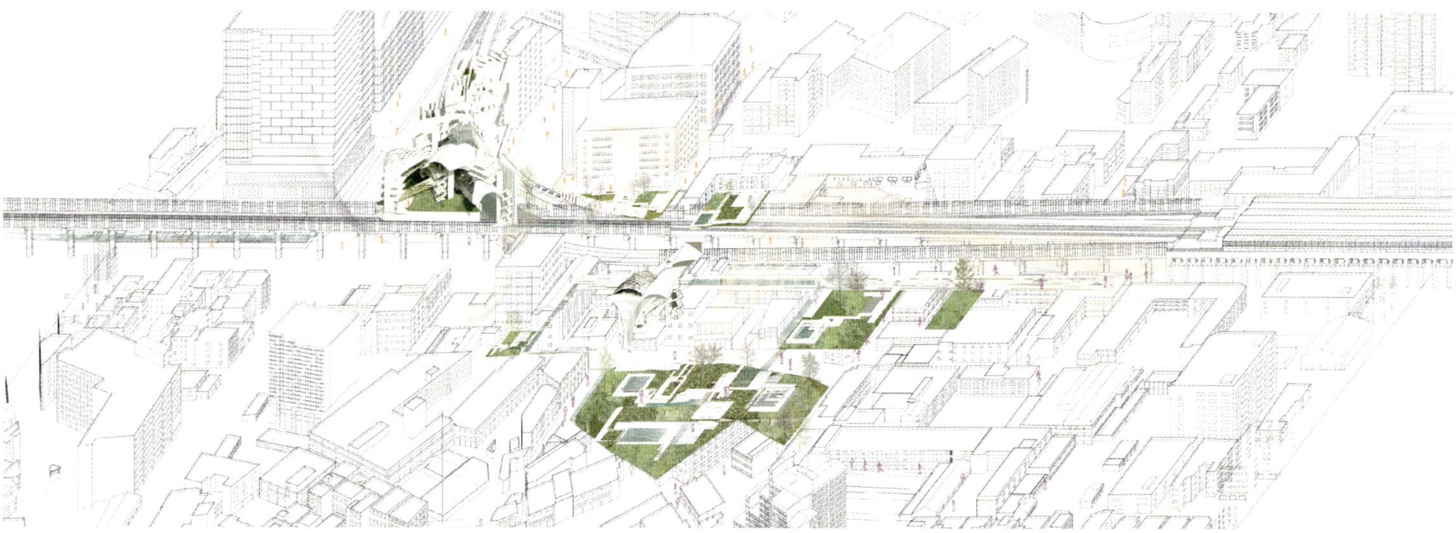

3

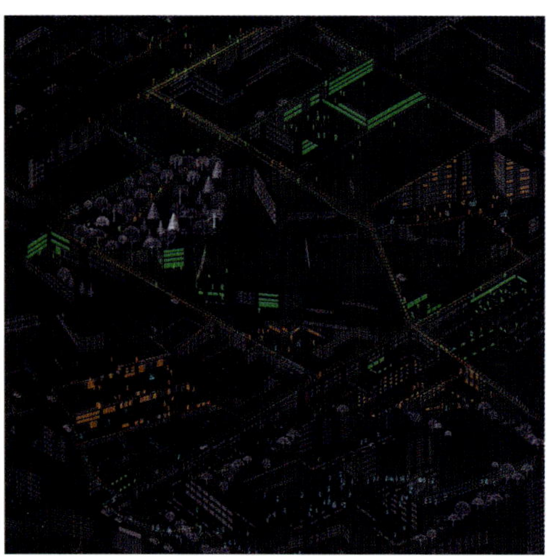

4. Intersected morning

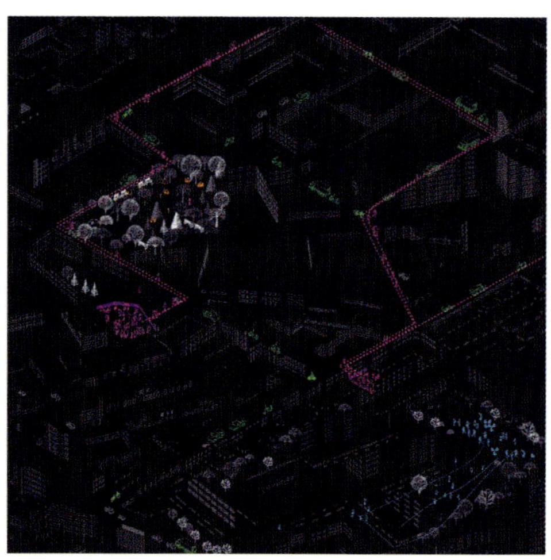

5. Separation at noon

6. Silent midnight

1-2. Concept Drawing

3. Masterplan

The project excavates the stream flow underneath the overpass, to create a water path and vapor garden that augments the existing fabric projecting many scenes along the way : these consist of not only delicate phenomena but also strategic program - bathhouse and residence – working as a bridge linking people's relationship.

4-7. Circulation Diagram & Exploded Axon

A shell roof structure providing the overall form is identified by an analysis on the wind path as the favorable geometry in terms of ventilation. Programs are arranged considering different time zones, each of which bears its respective circulation of the locals, factory workers, visitors of cafeteria.

7

8. Water Path
Before subway was built, the boundaries of the river were flexible and field was divided irregularly. The project permeates into existing urban fabric by excavating streams running under the overpass and creating new waterways.

8

9

10

11

9-17.
The project owes its name to its rich spatial experiences, which are found inside transitioning circumstances of everyday life. Firstly, the car glass reflected on concrete pillars of the overpass bridge, creates a wave-like vision, which in turn is enjoyed by people resting in underground baths. Secondly, as the subway passes, people aboard may experience the contrast of light and shadow created by overlapping shell structures. Finally, the transition under the form of a vague boundary created by the vapor garden, takes place.

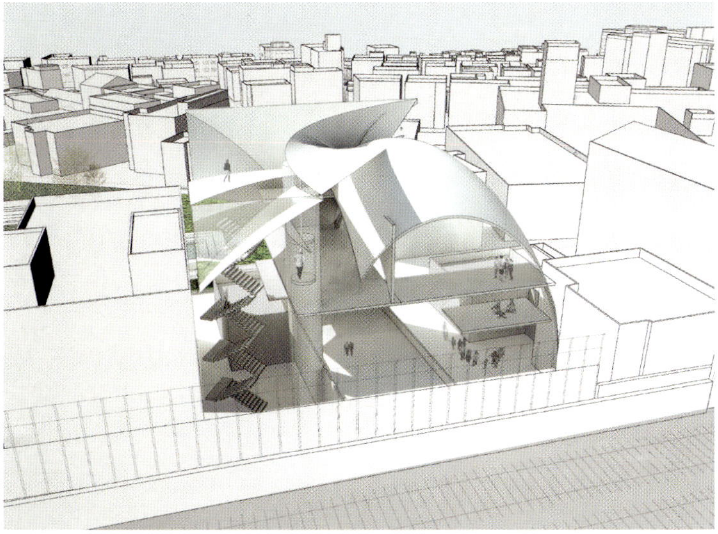
12. Open studio

13. Residence

14. Pocket park

15. Water path

16.Bath house

17. Vapor garden

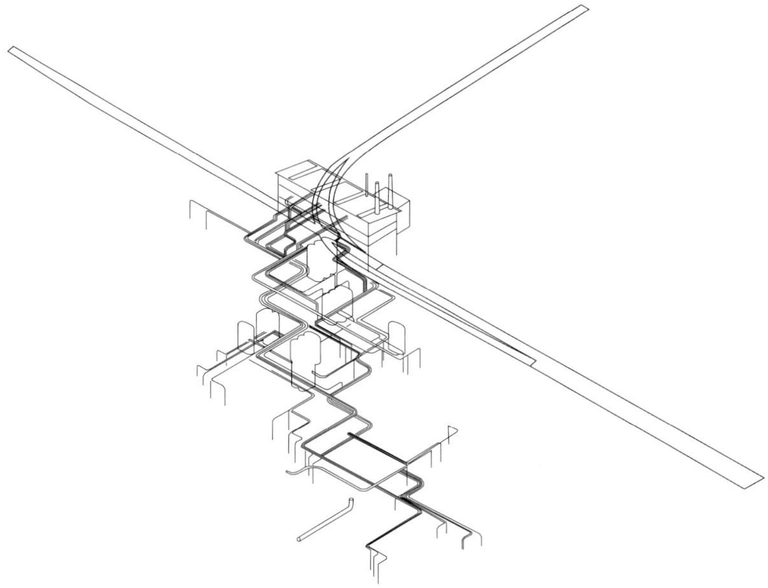

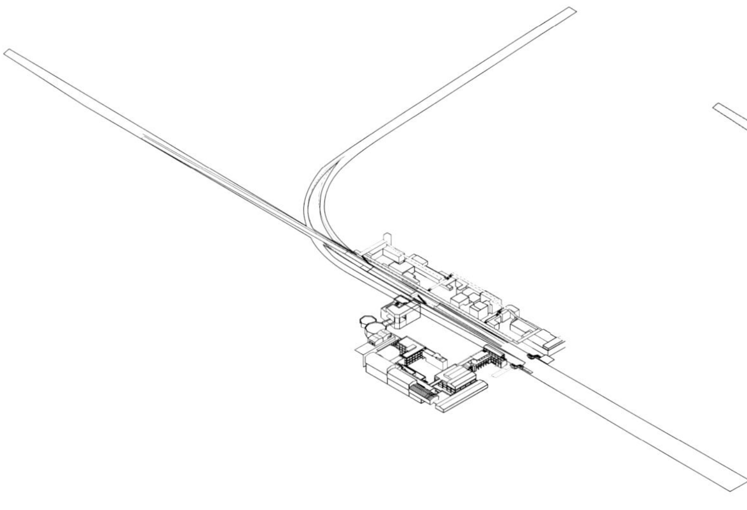

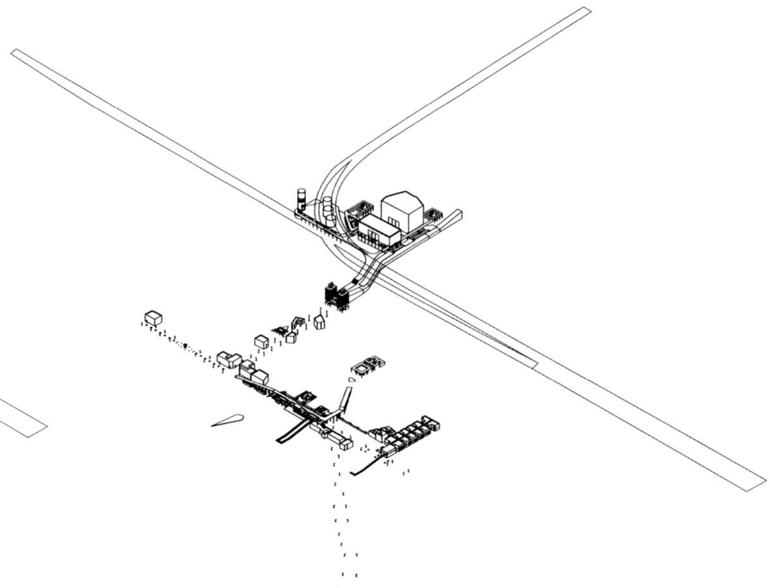

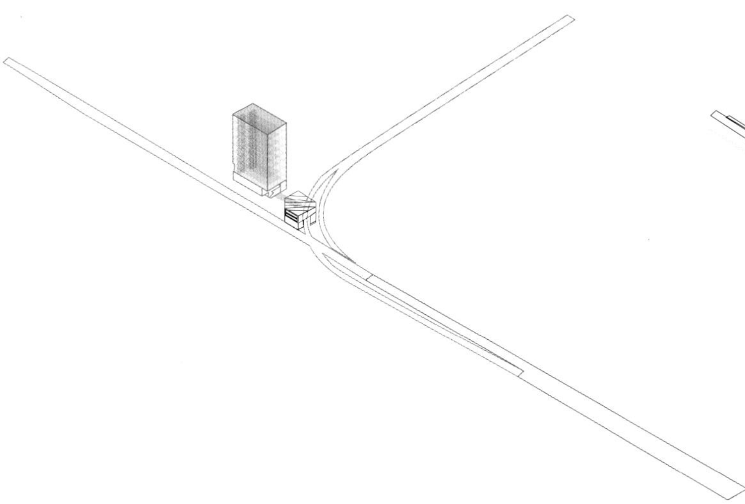

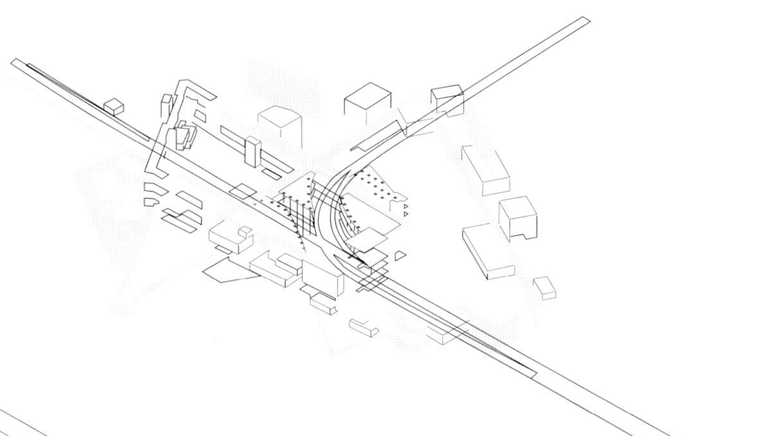

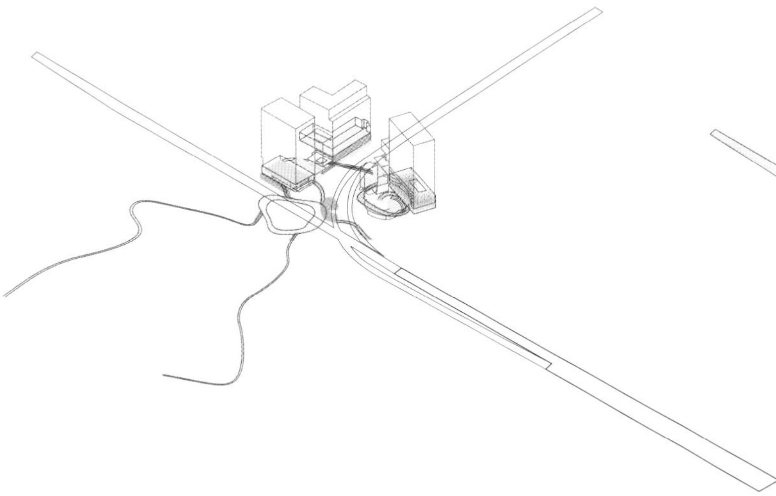

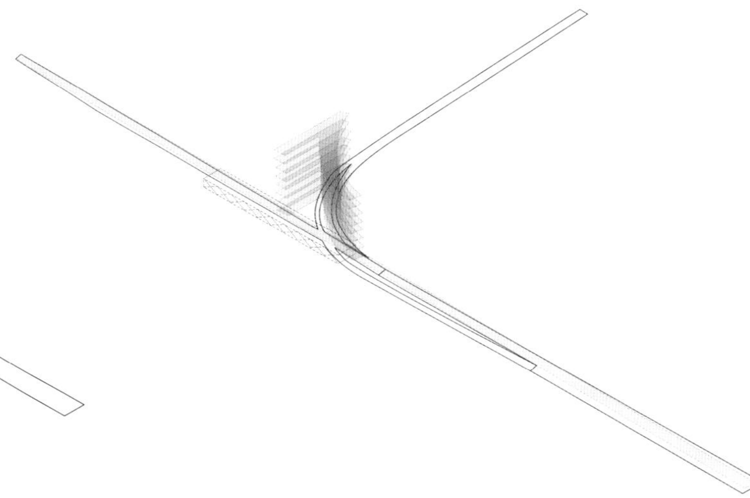

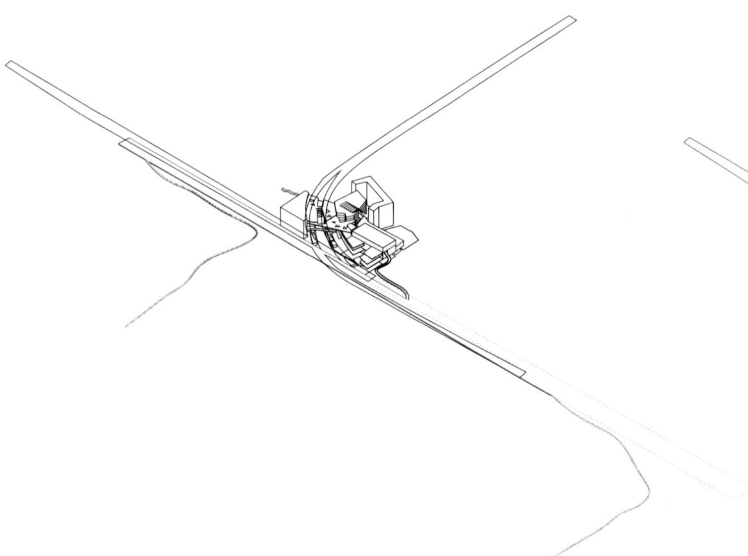

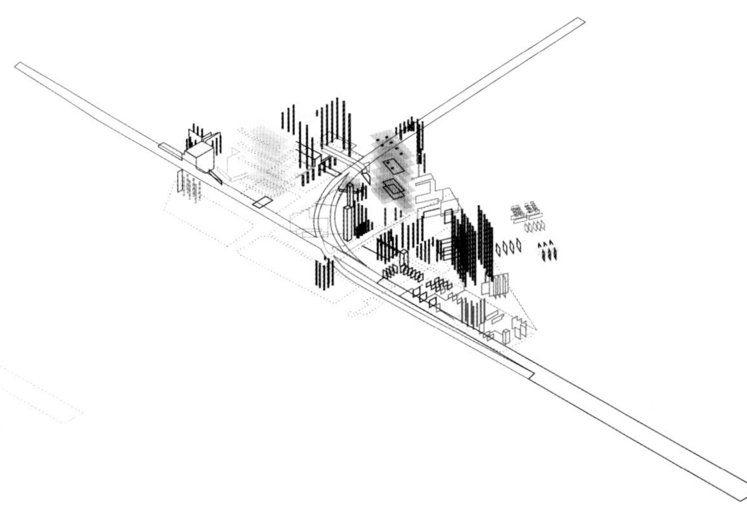

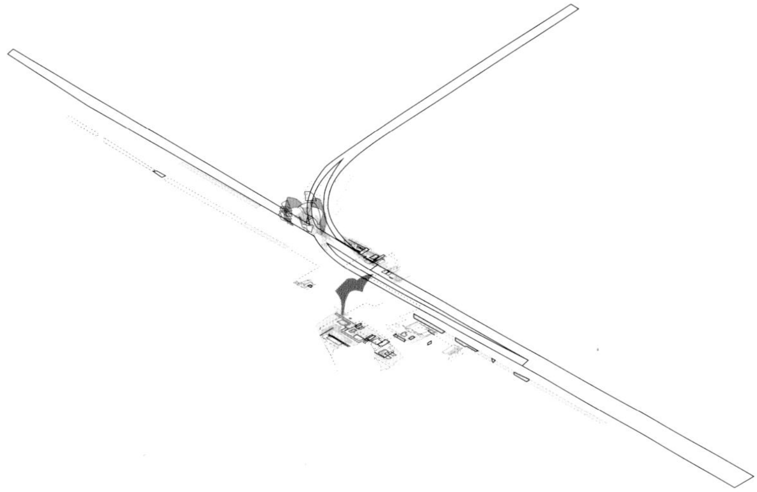

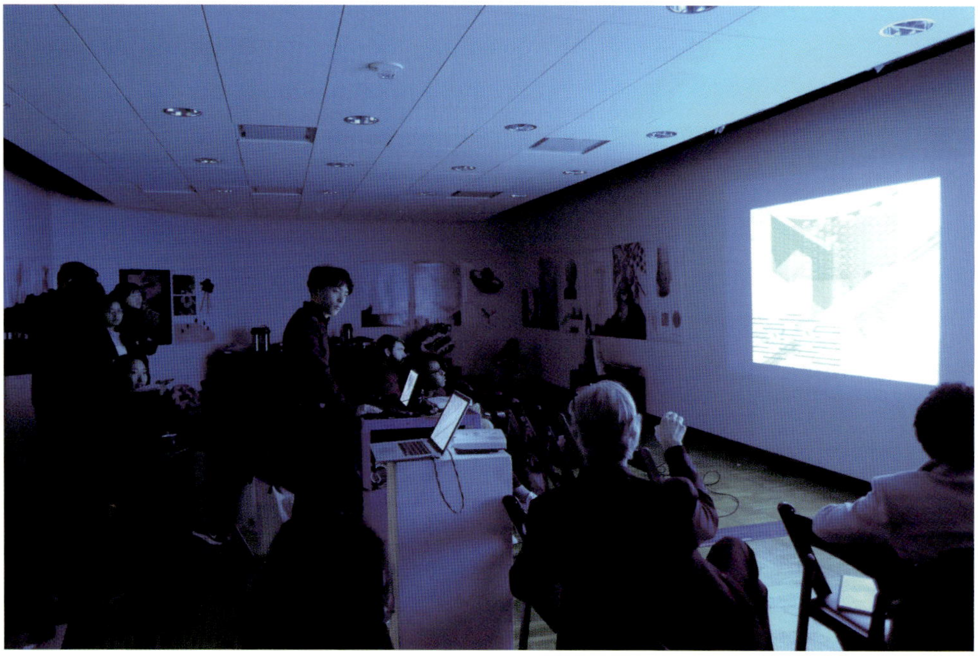

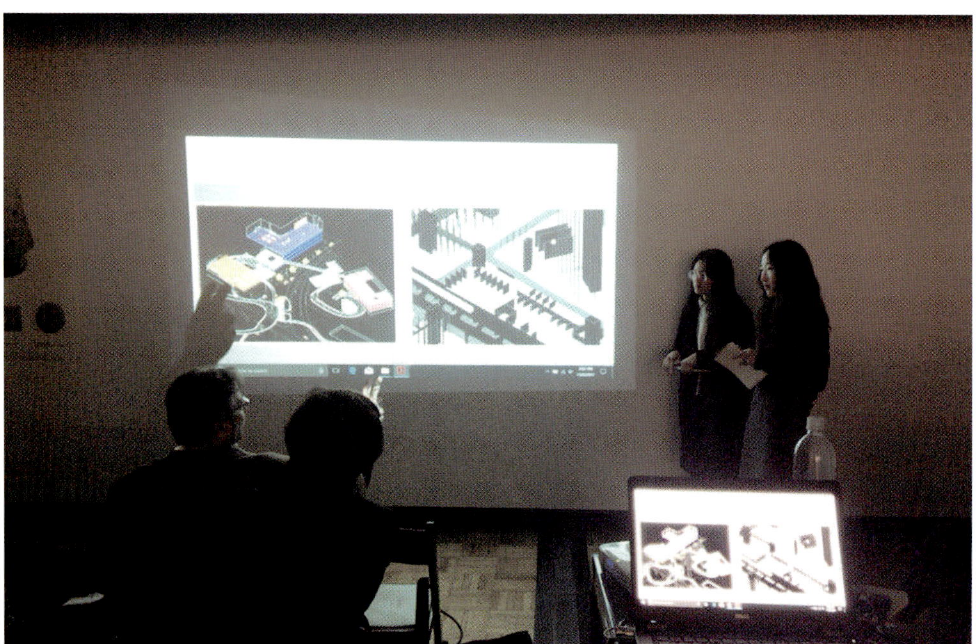

초판 1쇄 2018년 6월 8일

저자: John Hong
발행인: 김기현
발행처: 시공문화사(***Spacetime***)
출판등록: 1993년 3월 12일
주소: 서울시 서대문구 독립문공원길 13(03733) 극동프라자 5층
전화: 02) 3147-1212
팩스: 02) 3147-2626

http://www.spacetime.co.kr
mail: spacetime@korea.com

인쇄: (주)서울문화사
제본: (주)서정바인텍
용지: (주)대림지업사

ISBN: 978-89-5592-400-8

정가: 12,000원